Field & Laboratory Exercises

IN ENVIRONMENTAL SCIENCE

EIGHTH EDITION

Eldon D. Enger
Delta College

Bradley F. Smith
Western Washington University

Karen E. Lionberger
University of Georgia

Mc
Graw
Hill

Connect
Learn
Succeed™

The McGraw·Hill Companies

Mc Graw Hill

Connect
Learn
Succeed™

FIELD & LABORATORY EXERCISES IN ENVIRONMENTAL SCIENCE, EIGHTH EDITION

Published by McGraw-Hill, a business unit of The McGraw-Hill Companies, Inc., 1221 Avenue of the Americas, New York, NY 10020. Copyright © 2013 by The McGraw-Hill Companies, Inc. All rights reserved. Printed in the United States of America. Previous editions © 2000, 1997, and 1995. No part of this publication may be reproduced or distributed in any form or by any means, or stored in a database or retrieval system, without the prior written consent of The McGraw-Hill Companies, Inc., including, but not limited to, in any network or other electronic storage or transmission, or broadcast for distance learning.

Some ancillaries, including electronic and print components, may not be available to customers outside the United States.

This book is printed on acid-free paper.

8 9 0 LKV 24 23 22

ISBN 978—0—07—759982—9
MHID 0—07—759982—9

Senior Vice President, Products & Markets: *Kurt L. Strand*
Vice President, General Manager, Products & Markets: *Marty Lange*
Vice President, Content Production & Technology Services: *Kimberly Meriwether David*
Managing Director: *Thomas Timp*
Brand Manager: *Michelle Vogler*
Marketing Manager: *Matthew Garcia*
Content Project Manager: *Robin Reed*
Senior Buyer: *Sandy Ludovissy*
Designer: *Tara McDermott*
Compositor: *S4Carlisle Publishing Services*
Printer: *LSC Communications*

All credits appearing on page or at the end of the book are considered to be an extension of the copyright page.

The Internet addresses listed in the text were accurate at the time of publication. The inclusion of a website does not indicate an endorsement by the authors or McGraw-Hill, and McGraw-Hill does not guarantee the accuracy of the information presented at these sites.

Some of the laboratory experiments included in this text may be hazardous if materials are handled improperly or if procedures are conducted incorrectly. Safety precautions are necessary when you are working with chemicals, glass test tubes, hot water baths, sharp instruments, and the like, or for any procedures that generally require caution. Your school may have set regulations regarding safety procedures that your instructor will explain to you. Should you have any problems with materials or procedures, please ask your instructor for help.

CONTENTS

PREFACE

Major Revision

The purpose of this laboratory and field manual is to provide students with hands-on experiences that help them put theory into practice and develop a better understanding of the process of science and the tools used by environmental scientists. This eighth edition of *Field and Laboratory Exercises in Environmental Science* is essentially a new work. Sixteen of the twenty-two exercises are new and the others have been extensively revised. The twenty-two exercises include those that highlight ecological principles, population issues, environmental assessment, plate tectonics, pollution effects, energy topics, solid waste issues, climate change, and land-use planning. There are also two new appendixes. One provides guidelines for planning an effective field trip and the other provides suggestions for alternative learning activities and field trips.

Since the activities in this manual illustrate foundational concepts of any environmental science course, these exercises can be used in conjunction with any introductory environmental textbook. The exercises have been used by the authors in their teaching and have been tested for clarity and effectiveness.

New Coauthor

Karen E. Lionberger is a new coauthor for the 8th edition of *Field and Laboratory Exercises in Environmental Science*.

Karen has been actively writing in the field of environmental science for over six years. Her early research interests were in oceanic ecosystems and she earned a B.S. degree in marine biology from the University of North Carolina in Wilmington. Her interests evolved to include development of instructional and laboratory curricula for introductory college environmental science courses. Most recently, she has been working with higher education faculty from across the country to produce effective curriculum for introductory environmental science courses. She is currently a doctoral student at the University of Georgia.

Pedagogy

The exercises provide students with experiences that highlight key concepts in environmental science and allow students to foster critical thinking about environmental issues. The exercises provide opportunities for students to experience the process of science by engaging in posing questions, defining and refining hypotheses, and analyzing data. Since most scientific investigations involve a review of current knowledge on a topic, several of the exercises make use of data sets available from government agencies and other sources. Other exercises focus on typical methods and techniques actually used by environmental scientists to gather data.

Each exercise begins with a statement of the purpose of the exercise and a list of objectives that provide students with a clear understanding of what they are expected to gain from the investigation. Each exercise has an introduction with background material that helps students see how the investigations within each exercise relate to the environmental principles they have studied in their textbook. The procedures section provides guidance for the investigation and data collection. An important part of every exercise is the data analysis section. This section is intended to cultivate the student's ability to appropriately represent data in a way that will allow them to analyze their results and draw conclusions. Each exercise contains a set of data sheets that helps students organize their data, allows them to use graphs and other analysis techniques, and provides questions that require students to think about their results and what they mean in relation to environmental principles and concepts. Therefore, many of the questions at the end of each exercise require students to generalize their thinking from the specific activities of the exercise to broader issues in environmental science.

Response to Reviewer Suggestions

With the 8th edition of *Field and Laboratory Exercises in Environmental Science* we benefited from the input of reviewers and responded to several common themes. Reviewers strongly supported the inclusion of accurate data collection, the use of various tools to analyze data, and that students be able to support their conclusions. The data collection, analysis, and review questions at the end of each exercise accomplish this goal.

Reviewers also suggested that the manual include exercises that address specific environmental topics. The list of suggested topics and the exercise that address the requests follow:
1. Effects of pollution on living organisms
   - Exercise 10. Stream Ecology and Assessment
   - Exercise 13. The Effects of Radiation on the Germination and Growth of Squash Seeds
   - Exercise 15. Toxicity Testing (LD_{50})
   - Exercise 16. Effects of Salinization on Plants
2. Global climate change
   - Exercise 19. Global Indicators of Climate Change
3. Performing an ecological footprint calculation
   - Exercise 20. Evaluating Ecological Footprint Calculations
4. Ecological relationships and trophic levels
   - Exercise 1. Primary Productivity
   - Exercise 2. Habitat and Niche
   - Exercise 3. Community Structure
   - Exercise 10. Stream Ecology and Assessment
5. The importance of native plants and biodiversity
   - Exercise 2. Habitat and Niche
   - Exercise 3. Community Structure
6. Carcinogens, teratogens, and mutagens
   - Exercise 13. The Effects of Radiation on the Germination and Growth of Squash Seeds
   - Exercise 15. Toxicity Testing (LD_{50})
7. Land-Use planning
   - Exercise 21. Land-use Planning on Campus
8. Use of government data bases
   - Exercise 8. Plate Tectonics
   - Exercise 14. Evaluating Renewable Energy Sources
   - Exercise 18. Air Pollution
   - Exercise 19. Global Indicators of Climate Change
   - Exercise 20. Evaluating Ecological Footprint Calculations

Field and Laboratory Exercises in Environmental Science, 8th Edition
Detailed List of Changes

Exercise 1. Primary Productivity

This is a new exercise.

- This exercise uses the growth of aquatic plants in conjunction with oxygen test kits to measure the amount of oxygen in the water of plants that are in the dark and in light. This information allows students to calculate the gross primary productivity and the net primary productivity.
- It also includes a section on the effect of temperature on the amount of oxygen dissolved in water.

Exercise 2. Habitat and Niche

This is a modification of Exercise 5 from the previous edition.

- The purpose, objectives, and introduction were rewritten.
- The exercise looks at 5 plant species. The previous edition looked at 3.
- The part of the exercise that deals with site characterization is now organized as a table that focuses on soil type, soil moisture, slope, sunlight, and degree of human disturbance. This makes it easier for students to collect information on the character of the sites.
- A blank graph is provided for students to graph the abundance of each of the 5 species in each of the 3 habitats.
- The questions at the end of the exercise are new.

Exercise 3. Community Structure

This is a modification of Exercise 2 from the previous edition.

- The purpose, objectives, and introduction were rewritten.
- The directions for the exercise were reorganized to make them easier to follow.
- The data sheets now provide a table and grid map for the students to complete for each of the two quadrats.
- The instructions and tables that deal with relative density and relative frequency calculations have been reorganized and the tables have been reorganized to make it easier for the students to complete the calculations.
- New questions were written to include analysis of the distribution (clumped, random, even) and the vegetation maps the students constructed.

Exercise 4. Estimating Population Size

This exercise is a modification of Exercise 3 from the previous edition.

- The purpose, objectives, and the introduction were rewritten.
- The exercise provides three options for how the exercise can be completed, one of which involves marking mealworms in the lab. The other two options involve instructors selecting organisms and capture techniques.
- The analysis questions are new.

Exercise 5. Population Dynamics

This is a new exercise.

- The exercise uses the growth of a population of duckweed to explore the differences in population growth patterns between K- and r-strategists.
- Students calculate the biotic potential of duckweed.
- The exercise allows students to form a hypothesis about the population growth pattern of duckweed and test it.

Exercise 6. Historical Changes in Human Population Characteristics

This exercise is a modification of Exercise 9 from the previous edition.

- The purpose, objectives, and introduction were revised.
- The procedure section was revised.
- The questions are new.

Exercise 7. Human Population Dynamics

This is a new exercise.

- The exercise uses data from the Population Reference Bureau website.
- This exercise explores social and economic characteristics and demographic and environmental impacts of countries at different stages of the demographic transition.
- Students calculate the future population size of a country based on current demographic information.
- Students are asked to evaluate the effect of various demographic characteristics on future population growth.

Exercise 8. Plate Tectonics

This is a new exercise.

- This exercise uses data from the U.S. Geological Survey website.
- The introduction introduces students to the concept of plate tectonics.
- Students complete a mapping activity that shows the location of the 17 strongest earthquakes and relates their location to plate boundaries.
- Students also map the location of examples of specific kinds of plate boundaries.

Exercise 9. Soil Characteristics and Plant Growth

This is a new exercise.

- This exercise evaluates the kind of soil (gravel, sand, clay, potting soil) on the growth of plants.
- Students measure the water-holding capacity of the various soil types.
- Students evaluate the effect of fertilizer on plant growth.
- Students measure root length and plant biomass to determine the effect of soil type and the effect of fertilizer on plant growth.

Exercise 10. Stream Ecology and Assessment

This is a new exercise that incorporates some of the ideas of Exercises 13 and 14 from the previous edition.

- Students collect water quality data from two different streams—one with good water quality and one with poor water quality.
- The following water quality data are collected: stream velocity, temperature, pH, total suspended solids, dissolved oxygen, total soluble phosphates, nitrogen compounds, benthic macroinvertebrates, standard plate count of bacteria, and coliform bacteria count.
- The data collected are used to compare the water quality of the two streams studied.

Exercise 11. Personal Energy Consumption

This exercise is a modification of Exercise 20 from the previous edition.

- The purpose, objectives, and introduction were rewritten.
- In the procedure section there is a new section that allows students to calculate the annual heat loss/gain.
- The section on lighting in the procedure section now includes halogen and LED light bulbs.
- In the procedure section there is a new part that allows the student to calculate the long-term energy costs of appliances.
- The tables on the data sheets were revised.
- The questions are new.

Exercise 12. Insulating Properties of Building Materials

This is a modification of Exercise 19 from the previous edition.

- The purpose, objective, and introduction were revised.
- A new table on the data sheet makes it easier for students to record data.
- The questions were revised and several are new.

Exercise 13. The Effects of Radiation on the Germination and Growth of Squash Seeds

This is a new exercise.

- This exercise uses squash seeds exposed to different amounts of radiation to assess the effect of radiation on germination and growth.
- Students are asked to determine threshold doses.
- Students graph the effect of dosage on germination and growth.

Exercise 14. Evaluating Renewable Energy Sources

This is a new exercise.

- This exercise uses data from the National Renewable Energy Laboratory website.
- Students are asked to evaluate the feasibility of wind energy development for their locality.
- Students calculate the potential electrical power from photovoltaic collectors for their locality.
- Students compare the price per unit of energy for gasoline and ethanol.
- Students evaluate the potential for biomass energy development for their locality.
- Students evaluate the potential for commercial grade geothermal energy development for their locality.
- Students use a levelized cost of energy analysis to compare renewable energy sources with traditional fossil fuel and nuclear energy sources.

Exercise 15. Toxicity Testing (LD$_{50}$)
    This is a new exercise.
- This exercise uses brine shrimp to determine the LD$_{50}$ concentration of several common toxic materials found in the home.
- Students collect data on the number of brine shrimp that die at different concentration of a toxic material and graph the results to determine an LD$_{50}$ concentration.

Exercise 16. Effects of Salinization on Plants
    This is a new exercise.
- This exercise uses plants grown in soils exposed to different concentrations of salt to evaluate the effect of salinization on plant growth.
- Students graph the height of plants vs. salt concentration.
- Students are asked to relate salinization to irrigation practices.

Exercise 17. Dissolved Oxygen and Biochemical Oxygen Demand
    This is a new exercise.
- This exercise uses dissolved oxygen test kits to determine the biochemical oxygen demand of several water sources.
- Students also use a graphic to determine percent saturation of oxygen.
- Students can compare the water quality of different sources of water by comparing the biochemical oxygen demand.
- Students are asked to relate temperature, time of day, kind of water source, and season of year to the oxygen concentration and BOD of the water.

Exercise 18. Air Pollution
    This is a new exercise.
- This exercise uses standard tests for carbon monoxide, carbon dioxide, nitric oxide, and sulfur dioxide to measure the gases coming from automobile exhaust.
- Students perform the air pollution tests, graph the results, and evaluate the pollution produced by different kinds of vehicles.
- Students measure the ozone concentration of the ambient air and relate it to the time of day and air temperature.

Exercise 19. Global Indicators of Climate Change
    This is a new exercise.
- This exercise uses data from the Environmental Protection Agency, National Aeronautics and Space Administration, and National Oceanic and Atmospheric Administration to evaluate the potential consequences of climate change.
- Students are asked to plot change in carbon dioxide, nitrous oxide, and methane emissions.
- Students are asked to state the sources of various greenhouse gases.
- Students are asked to state trends in carbon dioxide concentration, global surface temperature, changes in arctic sea ice and land ice, and changes in sea level and state the consequences of a continuation of the trends.
- Students are asked to calculate their greenhouse gas footprint and determine ways to reduce their footprint.

Exercise 20. Evaluating Ecological Footprint Calculations

This is a new exercise.

- This exercise asks students to access two ecological footprint calculators on the Internet and evaluate their thoroughness.
- Students are asked to evaluate why certain types of questions are included in ecological footprint calculations.
- Students are asked to make theoretical changes to their lifestyles and use ecological footprint calculators to evaluate the effectiveness of their theoretical lifestyle changes.

Exercise 21. Land-Use Planning on Campus

This is a new exercise.

- Students are asked to use their campus as a framework for applying land-use principles.
- Students will collect resources (maps, land-use plans) and interview campus officials in charge of land-use practices and decisions.
- Students use a checklist to evaluate the current land-use practices on campus.
- Following a class discussion of all the resources gathered, the class will develop a list of proposals for changes to land use, with attention to rationale, practicality, economics, and the influences of external forces.

Exercise 22. Solid Waste Assessment

This is a new exercise.

- Students are asked to collect all the waste they produce during a three-day period.
- The waste will be separated into categories and the amounts quantified by item and total weight.
- Materials that are to be recycled are inventoried separately from other waste.
- Students are asked to evaluate how they could reduce the amount of waste they produce.

Eldon D. Enger
Bradley F. Smith
Karen E. Lionberger

EXERCISE 1
PRIMARY PRODUCTIVITY

Purpose & Objectives

This exercise will examine the effects of temperature on the concentration of dissolved oxygen in aquatic ecosystems. Students will also measure oxygen concentrations present in water samples to determine the primary productivity and respiration rates of a common aquatic plant. After completing this exercise, the student should able to:

1. Describe how temperature affects the amount of dissolved oxygen into an aquatic system.
2. Analyze impacts human activities have on the temperature of aquatic ecosystems.
3. Measure oxygen concentrations to calculate the net primary productivity of aquatic producers.
4. Explain why net primary productivity is vital to the health of the ecosystem.

Introduction

The sun serves as the primary source of energy input for most of Earth's ecosystems. Through the process of photosynthesis, photosynthetic organisms (plants and phytoplankton) are able to trap sunlight energy and convert it into the energy in the chemical bonds of organic molecules such as sugar.

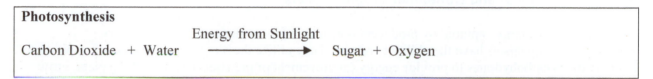

Photosynthesis

Carbon Dioxide + Water $\xrightarrow{\text{Energy from Sunlight}}$ Sugar + Oxygen

Thus, they are known as **producers.** The formation of organic compounds from carbon dioxide, water, and solar radiation is called **primary production**. The term "production" refers to the creation of biomass from the products of photosynthesis by producers (autotrophs). Producers constitute the first trophic level in the food chain. Each organism in an ecosystem belongs to a specific trophic level based on how it obtains nutrients. See Figure 1.1. All consumers are ultimately dependent on the biomass of producers. They either acquire their nutrients directly from eating producers, as in the case of primary consumers (herbivores), or indirectly, as in the case of secondary consumers (carnivores) that eat herbivores.

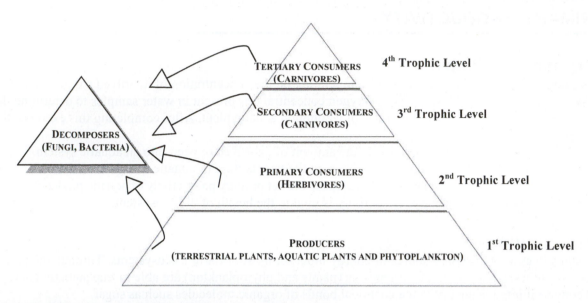

Figure 1.1 Producer and Consumer Trophic Levels

However, not all energy captured by producers is available for consumers. Like consumers, photosynthetic organisms have the ability to respire. They utilize the biochemical process of respiration to break down carbohydrates to provide energy for movement of materials throughout the plant, grow new plant parts, or respond in other ways to their environment.

Respiration
Sugar + Oxygen $\longrightarrow$ Carbon Dioxide + Water + Energy (ATP)

The difference between the amount of energy initially trapped by photosynthesis and the amount of energy remaining as plant biomass after plants have consumed some of the stored energy through respiration is called **net primary production**. Net primary production is the biomass available to consumers. The **gross primary production** of an ecosystem is simply the rate at which energy is captured during photosynthesis. If gross primary production and respiration rates are known, net primary productivity can be calculated as:

$$\text{net primary production} \quad = \quad \text{gross primary production} \quad - \quad \text{respiration}$$
$$\text{(plant growth)} \qquad\qquad \text{(total photosynthesis)}$$

In addition to providing biomass that serves as food for consumers, the process of photosynthesis releases oxygen used by plants, animals, and aerobic microorganisms during the process of respiration. In terrestrial ecosystems the availability of oxygen is rarely a limiting factor for consumers, since oxygen constitutes about 20% of the atmosphere. This is not the case for aquatic ecosystems. The amount of oxygen that can be dissolved in water is limited. The amount of oxygen present in water is determined by several factors. This includes:

- Photosynthesis by primary producers
- Respiration by producers and consumers
- Turbulence and wind at the air-water interface
- Temperature of the water

In this exercise you will use the amount of oxygen dissolved in water to assess the rate at which photosynthesis and respiration take place in aquatic plants. You will also determine how the temperature of the water affects the amount of oxygen dissolved.

Procedure

Materials Needed per Group of 3–4 Students

4 sealable glass bottles (either with rubber stoppers or screw caps)

Aquatic plants (*Elodea* or *Cabomba*)

Aged tap water

Aluminum foil

Fluorescent light

Dissolved oxygen probes or oxygen snap-tests or kits

250 mL beakers (4)

Thermometers

Water baths and/or incubators

Ice bath

Refrigerator

Labeling tape / marker

Part A:

1. Determine the oxygen concentration of the stock water that you will use as your source of water for this exercise. Record this information in all 4 boxes of the Dissolved Oxygen Initial row of Table 1.1 on Data Sheet 1.1.

2. Obtain 4 sealable bottles and label them as follows:

    Control Dark

    Control Light

    Plant Dark

    Plant Light

3. Fill the bottles with the stock water and treat as follows:

    a. Control Dark—Seal the bottle and wrap with aluminum foil so that no light enters the bottle.

    b. Control Light—Seal the bottle.

    c. Plant in Dark—Place 3 sprigs of your aquatic plant in the bottle, seal the bottle, and wrap with aluminum foil so that no light enters the bottle.

    d. Plant in Light—Place 3 sprigs of your aquatic plant in the bottle and seal the bottle.

4. Place all 4 bottles directly under a fluorescent light. The bottles should not be more than 30 cm away from the light source.

5.   You will need to wait approximately two hours before taking the final dissolved oxygen measurements. [Note: You should complete Part B of this procedure while you are waiting to take your measurements for Part A.]

6.   After at least two hours have passed, measure the concentration of dissolved oxygen in each of the bottles. Record your results in Table 1.1 on Data Sheet 1.1.

Part B:

1.   Prior to the lab, four 250 mL beakers were filled with tap water and placed in the following settings to achieve the approximate desired temperatures:

Sample	Desired Temperature
Ice Bath	0°C
Refrigerator	5°C
Lab Room	20°C
Incubator	25°C
Water Bath	30°C

2.   For each sample, use a thermometer to record the actual temperature and record the results in Table 1.2.

3.   Measure the amount of dissolved oxygen in each sample and record in Table 1.2.

4.   Plot the results in Table 1.2 on Graph 1.1 on Data Sheet 1.2. [Note: If a specific temperature was measured by more than one group, average the DO concentrations in order to plot it on the graph.]

EXERCISE 1
PRIMARY PRODUCTIVITY

Name:_____
Section:_____
Date:_____

Data Sheet 1.1

Table 1.1 Dissolved Oxygen Concentration

	Control in Light	Plant in Light	Control in Dark	Plant in Dark
Dissolved Oxygen Initial (mg/L)				
Dissolved Oxygen Final (mg/L)				

Table 1.2 Water Temperature and Corresponding Dissolved Oxygen Concentration

Sample	Temperature	Dissolved Oxygen Concentration
Ice Bath		
Refrigerator		
Lab Room		
Incubator		
Water Bath		

EXERCISE 1
PRIMARY PRODUCTIVITY

Name:_____

Section:_____

Date:_____

Data Sheet 1.2

Graph 1.1 Effect of Temperature on Dissolved Oxygen Concentration

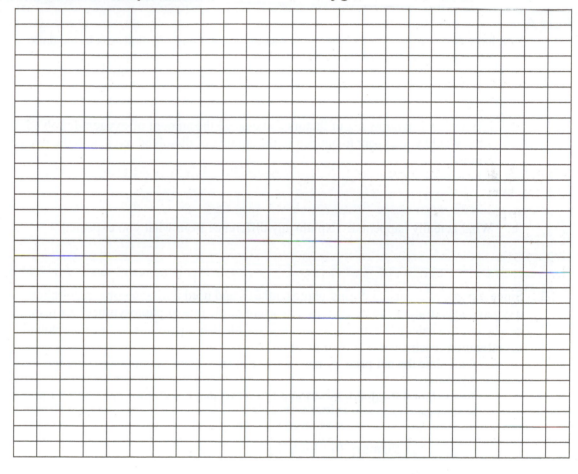

Dissolved Oxygen (mg/L)

Temperature (°C)

EXERCISE 1
PRIMARY PRODUCTIVITY

Name:_____

Section:_____

Date:_____

Data Sheet 1.3

Analysis

Part A

1. You have two control bottles—one that was in the light and one that was in the dark.
    a. What is the purpose of the two controls?

    b. From what you know about factors that change the amount of oxygen dissolved in water, what would you <u>expect</u> to happen to the amount of dissolved oxygen in the two control bottles?

    c. What actually happened? Explain any difference from what you expected.

2. Since oxygen is produced during photosynthesis and used during aerobic respiration, change in oxygen concentration can be used as a measure of the rate of photosynthesis or respiration and can be used to calculate primary productivity. Calculate the following for your aquatic plant investigation (each answer should be in mg/L).
    a. Calculate the net primary productivity in the "Light" sample by using the following formula:
        Net Primary Productivity = Final Light Oxygen Concentration – Initial Light Oxygen concentration

        Net Primary Productivity = _____

    b. Calculate the consumption of oxygen by respiration in the "Dark" sample by using the following formula:
        Respiration = Initial Dark Oxygen Concentration – Final Dark Oxygen concentration

        Respiration = _____

    c. Calculate the gross primary productivity in the "Light" sample by using the following formula:
        Gross Primary Productivity = Net Primary Productivity + Respiration

        Gross Primary Productivity = _____

EXERCISE 1
PRIMARY PRODUCTIVITY

Name:_____
Section:_____
Date:_____

Data Sheet 1.4

3. Explain why it was necessary to completely exclude light from the "Dark" bottle that contained plants.

4. Discuss the relationship between net primary productivity of producers and the available biomass in the ecosystem for consumers.

5. In a plant, over a 24-hour period (day and night) during the summer, will there be more photosynthesis than respiration, equal amounts of both, or more respiration than photosynthesis? Explain your choice.

6. Compare the net primary productivity of the plants in a body of water during the summer and winter. How will they be different and why?

7. Why might an agricultural scientist or a fisheries biologist want to know the gross primary productivity? Hint: they are both interested in getting the highest productivity from their ecosystems.

EXERCISE 1
PRIMARY PRODUCTIVITY

Name:_____
Section:_____
Date:_____

Data Sheet 1.5

Part B

8. Examine Graph 1.1. Describe the correlation you observed between increasing water temperature and dissolved oxygen concentration.

9. Many aquatic species, such as rainbow trout and striped bass, are sensitive to low levels of dissolved oxygen. Not only are these organisms vital as consumers in the ecosystem, they are also economically important to recreational fishing business. Discuss three human activities that could increase water temperatures in local aquatic ecosystems, such as streams and rivers, and therefore decrease the presence of these types of sensitive organisms.

Human Activity	Description of How This Activity Increases Water Temperature

10. In addition to water temperature, what other kinds of human activities could result in low oxygen concentrations in water?

Human Activity	Description of How This Activity Decreases Oxygen Concentration

EXERCISE 2
HABITAT AND NICHE

Purpose & Objectives

The purpose of this exercise is to develop an understanding of two important ideas in ecology: habitat and niche. Your instructor will select three habitats that vary significantly from one another and select five plant species that you will count in each of the three habitats. This will allow you to explore the concepts of habitat and niche by comparing the numbers of each plant species found in each of the habitats. After completing this exercise, the student will be able to:

1. Relate abiotic characteristics in a habitat to the kinds of plants found in the habitat.
2. Describe how species with a broad niche differ from those that have a narrow niche and provide examples from the species studied in the exercise.
3. Describe the concept of niche overlap and provide examples of niche overlap from the species studied in the exercise.
4. Identify biotic and abiotic environmental factors that may restrict the distribution of a species.

Introduction

The International Union for Conservation of Nature (IUCN) lists nearly 20,000 species as vulnerable or endangered. These species typically have narrow niches and specific habitat requirement. Therefore, habitat loss is a major cause of extinction. In order to protect endangered species it is necessary to have a good understanding of their niche characteristics and nature of the habitats they require.

An organism's **habitat** is the place an organism lives. Often general types of environments are referred to as habitats because they are associated with certain assemblages of plants, animals, and other organisms. For example, a sandy beach habitat would have specific organisms associated with it that might be absent or rare in a deciduous forest habitat, or a cold-water stream habitat. Each kind of habitat supports many kinds of organisms that must share common resources.

An organism's **niche** is the role an organism plays in its ecosystem. A niche is a very abstract concept that includes everything that affects a species of organism and all of the impacts the organism has on its surroundings. To clarify the distinction between an organism's habitat and niche let's look a couple of examples. The habitat of a bison is grassland. Its niche includes its role as a consumer of grass, being a host to parasites, and migrating to find food. Other niche characteristics include: feeding habits, breeding habits, competition with other grazing animals, and physiological constraints. The habitat of cattails is a wetland. Its niche includes: carrying on photosynthesis, providing hiding places for many kinds of animals, providing nesting sites for redwing blackbirds, and many other functions.

A **community** consists of many different species of organisms that interact with one another. Those with similar niches will compete with one another. A concept known as the **competitive exclusion principle** states that no two organisms can occupy the same niche at the same time. Species with the same niche requirements will be in intense competition with one another and this competition is harmful to both organisms. In some cases one of the two competing organisms may simply not be able to survive and will be excluded from the area. Natural selection also plays a role. When two species are in intense competition, natural selection may lead to one species being driven to extinction. Alternatively natural selection may cause the two species to evolve into slightly different niches, which reduces the intensity of competition. Thus, even organisms that appear to have similar niches actually have some important niche differences. This separation of organisms into distinct niches is known as **niche differentiation**. For example, there are many species of birds that feed on insects. However, they may not compete directly because they feed on different kinds of insects, look for insects in different places, or feed at different times of the day. Thus, niche differentiation allows many species of insect-eating birds to coexist without intense competition.

The range of conditions in which an organism can be successful is known as **niche breadth**. Some species can live in a wide variety of habitats and have very broad niches, while other species require very special sets of environmental conditions and have very narrow niches. For example, the American robin has a very broad niche. It eats many kinds of foods, can survive in a wide variety of environmental conditions, and is found throughout North America. However, the saguaro cactus has a very narrow niche. It requires summer rains for seeds to germinate, rocky soil to anchor its roots, and very mild winter temperatures. Thus, it is found in a very small area of the desert southwest where the proper environmental conditions occur.

Niche overlap is the degree to which different species play similar roles in their communities. All plants have a certain degree of niche overlap because they all carry on photosynthesis. All plants that have insect-pollinated flowers have a certain amount of niche overlap. However, there is also niche differentiation among plants that use different kinds of insects as pollinators. Similarly, the niches of wind-pollinated plants do not overlap with those of insect pollinated plants. The difference in method of pollination (niche differentiation) reduces niche overlap and reduces competition.

Procedure

Materials Needed for Each Group

4 pieces of surveyor's ribbon (about 18 inches long)

Colored pencils

50 meter tape measure

Two meter sticks

Carpenter's level

Data sheets

Calculators

Method
1. Your instructor will describe five plant species that are common in your area and that occur in at least one of the three habitats. Each species will be identified by a letter. Record the name of the species corresponding to each letter on Table 2.2 on Data Sheet 2.2.
2. Your instructor will divide the class into five groups. Each group will be assigned one of the five plant species, and will be responsible for counting only the plants of that species at each habitat.
3. Your instructor will give directions on where the three habitats are located.
4. At each of the three habitats, use a 50-meter tape to measure a 30-meter by 30-meter square plot and mark the corners with the surveyor's ribbon.
5. At each of the three habitats, students will characterize the site by gathering the data requested on Table 2.1 on Data Sheet 2.1 and entering the information in the table.
6. Each group is responsible for identifying and counting every plant of its assigned species within the 30 × 30 meter plot in each of the three habitats and recording the number counted on Table 2.2 on Data Sheet 2.2. If you are dealing with trees, do not count seedlings or saplings less than 5 centimeters in diameter. The instructor will assist students in identifying their species or provide resources for doing so.

7. After returning to the classroom, each group will share its results and all students will record these results on Table 2.2 Data Sheet 2.2 and calculate the average for each species.
8. Construct a graph of the number of each species on Graph 2.1 Data Sheet 2.2. Use different colors or shading to plot the numbers of each species (A, B, C, D, and E).
    a. The following example will help you determine how to plot the graph. See Figure 2.1 below.
        ▪ Species A is shown with black bars. Assume there were 12 individuals in Habitat 1, 22 individuals in Habitat 2, and 5 individuals in Habitat 3.
        ▪ Species B is shown with light gray bars. Assume there were 10 individuals in Habitat 1, 5 in Habitat 2, and 30 in Habitat 3.
        ▪ Species C is shown with dark gray bars. There were 8 individuals in Habitat 1, 12 in Habitat 2, and 3 in Habitat 3.
        ▪ We can also calculate the average of each of the three species across all habitats by adding up all the individuals of a species and dividing by 3. Graphically, the distribution would look like figure 2.1.
    b. By examining figure 2.1, we can evaluate niche breadth and overlap as follows:

Niche Breadth

To evaluate niche breadth, compare the number of individuals of each species in each of the three habitats with the average.

Species that deviate most from the average have narrow niches. In our example, Species B has a narrower niche than the other two species. If you look at the three sites, you can see that the majority of individuals of Species B are found in only one habitat. It might be termed a habitat "specialist."

If a species exists in all three habitats in about the same numbers, the average for that species will be close to the number found in each habitat. Species that have populations in all three habitats that are close to the average have broad niches. In our example, Species C occurs in about the same numbers in all three habitats. It is a "generalist."

Niche Overlap

To evaluate the degree to which the niches of the three species overlap in a habitat look at each habitat separately. The greatest degree of niche overlap between Species A, B, and C occurs in those habitats where the numbers of the three different species are similar. In this example, there is some degree of niche overlap among Species A, B, and C in all three habitats, with the greatest overlap in Habitat 1 and the least overlap in Habitat 3.

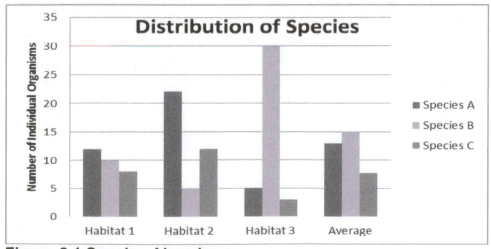

Figure 2.1 Species Abundance

EXERCISE 2
HABITAT AND NICHE

Name:_____
Section:_____
Date:_____

Data Sheet 2.1

Characterize each of the three habitats and record the data in the spaces provided.

Table 2.1 Site Characteristics

Site Characteristics	Habitat 1	Habitat 2	Habitat 3
Soil Type *Rocky*—pebbles, rocks, and stones common on the surface *Sandy*—most particles the size of grains of sugar *Loam*—mixture of particle sizes *Clay*—very fine particles (If dry, clay may form chunks.) *Organic*—pieces of partially decomposed plants evident			
Soil moisture *Dry*—dusty *Moist*—damp to the touch *Wet*—can wring drops of water from a handful of soil (Or use a soil moisture meter.)			
Slope (Place one end of a meter stick on the soil with the free end downslope. Use a level to assure that the meter stick is level. Use a second meter stick to measure the distance from the free end of the meter stick to the soil on the down slope end. The number of centimeters measured between the horizontal meter stick and the surface is the % slope.) Steep—15% or greater Moderate—2%–15% Flat—Less than 2%			
Sunlight *Open*—Sun reaches tall plants throughout the day *Partly shaded*—Tall plants will be in shadow 25–50% of the day *Shady*—Tall plants will be in shade over 50% of the day			
Degree of Human Disturbance *Little*—Humans visit the area but don't modify it. *Moderate*—Evidence that humans have altered the area (paths, cut trees, camp sites) *Severe*—Regular alteration (mowing, tilling, timber harvest)			
Other characteristics your instructor asks you to note.			

EXERCISE 2
HABITAT AND NICHE

Name:_____
Section:_____
Date:_____

Data Sheet 2.2

Group Number: _____
Species for which you are responsible: _____

Record the total number of individuals of your plant species found in the study plot of each of the three habitats. After returning to the classroom, enter the data collected by the other groups.

Table 2.2 Number of Plants Found in Each Habitat

Species	Species' Name	Habitat 1	Habitat 2	Habitat 3	Total for All 3 Habitats	Average
Species A						
Species B						
Species C						
Species D						
Species E						

Use the data from Table 2.2 to construct a graph of the distribution of Species A, B, C, D, and E and each species' average. You may wish to enter the data from Table 2.2 into an Excel spread sheet to construct the graph.

Graph 2.1 Distribution of Species in Three Different Habitats

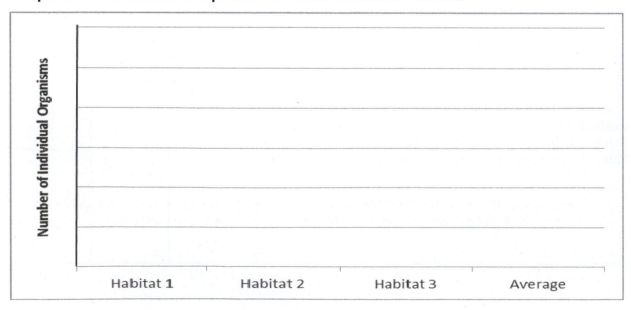

16

EXERCISE 2
HABITAT AND NICHE

Name:_____
Section:_____
Date:_____

Data Sheet 2.3

Analysis

Although there are statistical methods for evaluating the niche breadth of a species and the degree of niche overlap between species, they are somewhat cumbersome so we will evaluate niche breadth and niche overlap by examining the bar graph 2.1 of species abundance you produced. See the discussion of niche breadth and overlap at the beginning of this exercise.

1. Which of the 5 species examined were generalists? How do you know they are generalists?

2. Which of the 5 species examined were specialists? How do you know they are specialists?

3. For the specialist species, which of the environmental factors you assessed appeared to be most important to the success of the specialist species?

4. What characteristics of the generalist species appear to have accounted for their success in the different habitats?

5. In which habitat was the greatest amount of niche overlap observed?

6. Those species that occur together show some degree of niche overlap and are in competition with one another. What evidence do you have that there has been niche differentiation?

EXERCISE 3
COMMUNITY STRUCTURE

Purpose & Objectives

In this exercise we will use a quadrat method to analyze the structure of a plant community by identifying specific plant species and determining how common each species is in the study area. After completing this exercise the student will be able to:

1. Use the quadrat method for studying communities.
2. Determine the density and relative density of each species in the community.
3. Determine the frequency and relative frequency of each species in the community.
4. Determine which species of plants are dominant in a community.
5. Determine if plants have a clumped, random, or uniform distribution.

Introduction

A **community** consists of all the different species of organisms (plants, animals, and microorganisms) that occur in the same location and interact with one another in specific ways. Organisms of the same species that live in the same place constitute a **population**. Thus, a community consists of an assemblage of populations existing in a common area, which interact with each other and share the same general resources. An **ecosystem** is a broader concept than a community because in addition to the organisms (biotic components) it includes physical (abiotic) components. Abiotic components of the environment, such as precipitation, temperature, and soil help define the nature of a community. Some communities, such as the arctic tundra, are relatively simple, with only a few hundred species interacting. Others, like tropical rainforests, contain many thousands of species, which produce a complex web of interactions.

A complete study of **community structure** would include identifying all the organisms present and describing how they interact with one another. Because communities are incredibly diverse in structure and function, it is impractical to study every organism in a community. However, we can gain insight into the structure and function of a community by studying samples of a community.

In a community, it is usually easier to study the plants than animals or microbes, since plants don't move and are generally large enough to be identified easily. Although this exercise is designed to survey plants in a grassland, the exercise can be adapted to other kinds of communities (forests, deserts, or wetlands) by slightly changing the way species are counted. Your instructor will be able to help you determine which type of sampling method is most appropriate for your area.

Another aspect of community structure is the degree to which certain species are dominant members of the community. There are several ways to gain an idea of the **dominance** of a species. Two measures of dominance known as relative density and relative frequency compare numbers of organisms. However, these methods do not take into account the size of an organism. One large maple tree would have a larger influence on a community than one dandelion. Therefore, some way of evaluating size is important in determining dominance. In this exercise we will use determinations of the relative density, relative frequency, and the size of plants as we evaluate the degree to which certain species are dominant in their community.

Procedure

Materials
50 meter tape or meter stick
Several 1 m² quadrats
       (For habitat types other than grasslands different methods may be required.)
Marking flags
Notebook
Data sheets
Field guide of local grassland vegetation or other appropriated field guides

Method
(This exercise describes sampling in a grassland environment. If a grassland is not available in your area, other community types may be used with slight modification of procedure. Your instructor will give instructions on how the exercise is to be modified.)

1. Your instructor will assign students to groups of 2 to 4 students. Each group will count the plants in 2 quadrats in the field and share their data with the rest of the class when everybody returns to the lab.
2. Your instructor will provide examples: drawings, photographs, or descriptions of five major plant species found in the community you will study. Your instructor also will provide the letter designation and the name of the plant. Record the name of each plant with its corresponding letter designation on Table 3.1 on Data Sheet 3.1 and Table 3.2 on Data Sheet 3.2.
3. In the field, establish a square study area. Use the measuring tape to measure the length of the sides of the study area and mark the corners with the marking flags. For a grassland a 100-square-meter plot, which measure 10 meters on a side, would be adequate. However, for forests or deserts your instructor may suggest a larger size for the study area.
4. To obtain statistically valid results at least 10 percent of the study area should be sampled using quadrats. (If you have a 100 m² study area, a minimum of ten 1 m² quadrats should be surveyed).
   i) Starting at any edge along the study plot, each group will determine the location of its 2 quadrats by a method chosen by the instructor.
   ii) For the first quadrat, identify, count, and record the number of plants belonging to each of the five species on Table 3.1 at the top of Data Sheet 3.1.

   While counting, you must make several decisions. To include a plant in the count, at least 50 percent of it must be in the quadrat. For bushes and grasses, you must make a decision as to what is an individual plant, since one plant may be bushy and may look like several plants, when in fact it is not (your instructor will help you make these decisions).

   iii) On Figure 3.1 Vegetation Map on Data Sheet 3.1 record the species, size, and location of each plant in the quadrat. Indicate a plant's size by drawing a circle proportional to the size of the plant in the quadrat. Identify the species by using the letter designation for the species. What you will end up with is a "map" of the quadrat, showing the relative size and number of the five species you are counting.
5. For your second quadrat repeat the procedures and record results on Table 3.2 and Figure 3.2 on Data Sheet 3.2.
6. Return to the lab and share your data with the rest of the class. Complete Table 3.3 Class Results with the data collected by all groups in the class.
7. Complete the calculations for relative density and relative frequency on Data Sheets 3.3 and 3.4.

EXERCISE 3
COMMUNITY STRUCTURE

Name:_____

Section:_____

Date:_____

Data Sheet 3.1

Table 3.1 Plants Identified in First Quadrat

Species Letter Designation	Species Name	Number of Plants in Quadrat
Species A		
Species B		
Species C		
Species D		
Species E		

Figure 3.1 Vegetation Map Use the letter designation to identify each species and draw a circle representing the relative size of each plant in the vegetation map below.

	25 cm	25 cm	25 cm	25 cm
25 cm				
25 cm				
25 cm				
25 cm				

EXERCISE 3
COMMUNITY STRUCTURE

Name:_____
Section:_____
Date:_____

Data Sheet 3.2

Table 3.2 Plants Identified in Second Quadrat

Species Letter Designation	Species Name	Number of Plants in Quadrat
Species A		
Species B		
Species C		
Species D		
Species E		

Figure 3.2 Vegetation Map Use the letter designation to identify each species and draw a circle representing the relative size of each plant in the vegetation map below.

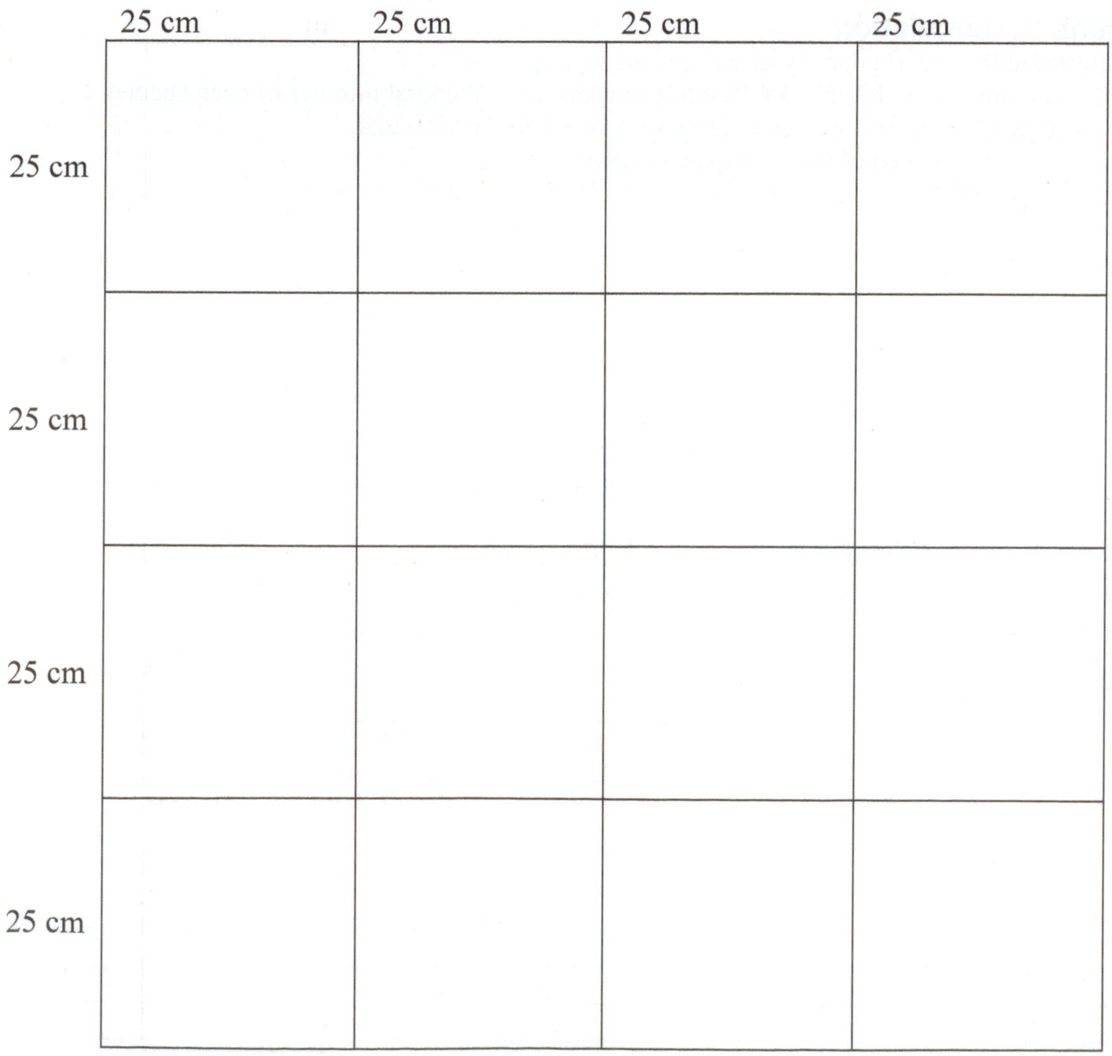

	25 cm	25 cm	25 cm	25 cm
25 cm				
25 cm				
25 cm				
25 cm				

EXERCISE 3
COMMUNITY STRUCTURE

Name:_____

Section:_____

Date:_____

Data Sheet 3.3

Table 3.3 Class Results

Quadrat Number	Number of Each Species in Each Quadrat														
	1	2	3	4	5	6	7	8	9	10	11	12	13	14	Total
Species A															
Species B															
Species C															
Species D															
Species E															
Total															

Relative Density Calculation

1. To determine the species density of each of the five species:
    a. In column (1) of Table 3.4 Relative Density enter the total number of each species counted from the far right column of Table 3.3 Class Results.
    b. In column (2) enter the total area sampled.
        (The total area sampled will be the size of the quadrat times the number of quadrats sampled by the class.)
    c. Divide column (1) total number of plants of *Species A* by column (2) total area sampled and record in column (3).
    d. Repeat for *Species B-E*.
2. To determine the total density of plants:
    a. Add the total of all the plants of all five species in all the quadrats in Table 3.3 Class Results. (the shaded box in Table 3.3)
    b. Divide the total number of plants by the total area sampled.
    c. Put this number in all the squares in column (4) Total Density in the Table 3.4 Relative Density below.
3. To determine the relative density of each of the five plants:
    a. Divide column (3) by column (4) and multiply by 100.
    b. Record results in column 5.

Table 3.4 Relative Density

Species	(1) Total Number of Plants	(2) Total Area Sampled (m^2)	(3) Species Density (Plants/m^2) (1) ÷ (2)	(4) Total Density (Plants/m^2)	(5) Relative Density (3) ÷ (4) × 100
A					
B					
C					
D					
E					

EXERCISE 3
COMMUNITY STRUCTURE

Name:_____
Section:_____
Date:_____

Data Sheet 3.4

Relative Frequency Calculation

1. To determine the frequency of each of the five kinds of plants:
   a. In column (1), for each species from Table 3.3 Class Results record the number of quadrats that contained that particular species.
   b. In column (2), record the total number of quadrats surveyed.
   c. Divide column (1) by (2) to get the frequency of each species. Record in column (3).
2. To determine the total frequency for all the quadrats sampled:
   a. Use the data in Table 3.3 Class Results to determine the number of quadrats that had any of the five plants included in the study.
   b. Divide the number of quadrats that had any of the five plants by the total number of quadrats sampled.
   c. Put this number in all the squares of column (4) Total Frequency in Table 3.5 below.
3. To determine the relative frequency of each plant species:
   a. Divide column (3) by column (4) and multiply by 100.
   b. Record results in column (5).

Table 3.5 Relative Frequency

Species	(1) Number of Quadrats in Which Species Occurs	(2) Total Number of Quadrats	(3) Frequency of Species (1) ÷ (2)	(4) Total Frequency	(5) Relative Frequency (3) ÷ (4) × 100
A					
B					
C					
D					
E					

EXERCISE 3
COMMUNITY STRUCTURE

Name:_____

Section:_____

Date:_____

Data Sheet 3.5

Analysis

1. A simple count of the number of plants present does not take into account the size of individual plants. Some plants may be very numerous but not occupy much space. In other cases a single plant may occupy a large space. The vegetation maps you created allow you to examine the impact of plants of different sizes.

    a. According to the vegetation maps you constructed, which species of plant appeared to be most dominant?

    b. Approximately what percent of the area of each quadrat was taken up by the most dominant plant?

    c. Did both of the quadrats you sampled show the same plant as being dominant?

    d. Compare your vegetation maps with other groups in the class. Did they have similar results? Why might they be different?

2. Based on the relative density of each species, which plant is the most dominant?

    a. Is this the same as the plant you identified as dominant in question 1?

    b. If they are different, what would account for the difference?

3. Based on the relative frequency of each species, which plant occurs most frequently?

    a. Is this the same species that was most dominant based on your evaluation of the vegetation maps and data on relative density?

    b. Explain any differences.

EXERCISE 3
COMMUNITY STRUCTURE

Name:_____
Section:_____
Date:_____

Data Sheet 3.6

4. Use the relative frequency of each species to determine if the distribution of each species is clumped, random, or uniform. Clumped species typically have a relative frequency between 0 and 30; randomly distributed species typically have a relative frequency between 31 and 80; and uniformly distributed species typically have a relative frequency between 80 and 100.

Species	Relative Frequency	Clumped/Random/Uniform
A		
B		
C		
D		
E		

a. List three characteristics of plants that might contribute to a clumped distribution.
   1.
   2.
   3.

b. What environmental characteristics might contribute to a uniform distribution of plants?

5. Are there any features or characteristics of the environment (e.g., slope, soil moisture, shade, etc.) that might lead to the distributions you observed?

6. What is the relation between density and frequency? Can a species of plant have a high relative density, but have a low relative frequency, or vice versa? Explain.

7. What difficulties could you encounter in trying to run quadrat studies on animal populations? In an aquatic environment?

EXERCISE 4
ESTIMATING POPULATION SIZE

Purpose & Objectives
This exercise explores a common method used by biologists to estimate population size. After completing this exercise the student will be able to:
1. Estimate the size of a population using the mark-recapture technique.
2. State factors that limit the accuracy of the mark-recapture technique.
3. State why estimates of populations may be necessary.

Introduction
Plants are relatively easy to census, since they stay in one place. However, animals move about, which makes it difficult to determine if each animal you see is a different individual. Furthermore, many animals are inconspicuous or active at night, which makes them difficult to count. Finally, many larger animals like bears or cougars have low population densities, which makes it very difficult to see or capture animals and identify individuals by visual sightings.

There are many reasons biologists want to know the size of a population. Game managers need to set hunting regulations that maintain an optimal population size. Biologists that monitor populations of endangered species need to know if a population is increasing, decreasing, or stable. Populations of parasites or other pest species often need to be controlled to maintain good crop yields.

One technique used to estimate population size is called a capture-recapture or mark-recapture method With this technique a number of individual organisms are captured and marked in some way. They are then released back into the population. The population is then sampled a second time and the number of marked and unmarked individuals are recorded. The ratio of initially marked individuals to the total population is equal to the ratio of marked individuals in the second sample to the total number captured in the second sample. Mathematically it looks like this.

$$\frac{\text{Total Population}}{\text{Number Marked Initially}} = \frac{\text{Total Individuals in 2}^{nd}\text{ Sample}}{\text{Number of Marked Individuals in 2}^{nd}\text{ Sample}}$$

This ratio can be rearranged as follows:

$$\text{Total Population} = \frac{\text{Total Individuals in 2}^{nd}\text{ Sample}}{\text{Number of Marked Individuals in 2}^{nd}\text{ Sample}} \times \text{Number Marked Initially}$$

For example, suppose you wanted to know the size of a population of monarch butterflies in a field.
1. You use an insect net to capture 10 monarchs and place a small dot of nontoxic paint on the abdomen of each butterfly.
2. You then release all ten monarchs near where you captured them.
3. A few days later you capture 14 monarch butterflies of which 7 are marked with paint.
4. You can then calculate an estimate of the total population as follows:

$$\frac{\text{Total}}{\text{Population}} = \frac{\text{Total Individuals in 2}^{nd}\text{ Sample } \textbf{\textit{(14)}}}{\text{Number of Marked Individuals in 2}^{nd}\text{ Sample } \textbf{\textit{(7)}}} \times \text{Number Marked Initially } \textbf{\textit{(10)}}$$

$$\text{Total Population} = \frac{14}{7} \times 10 = 20$$

This technique involves several assumptions.

1. The technique used to capture animals must sample animals randomly.
2. The animals that were captured, marked, and released must not behave differently from other animals.
3. Marked individuals must have time to mix freely before a second sample is taken.
4. Marked individuals must not lose their marks.
5. There must not be major changes in the population (births, deaths, migration, etc.) between the first and second samples.

If any of these assumptions is not met, the estimate is less accurate.

Procedure

There are several options for how this exercise can be conducted.

1. The exercise can be done in the lab with a population of meal worms. Meal worms can be raised easily in plastic containers containing wheat bran or chicken mash for food. Your instructor will provide several containers of meal worms and assign groups of students to work with each container.
    a. You will sift through the food, collect meal worms, mark them with indelible ink, and release them back into the plastic container.
    b. Record the number of meal worms marked on Table 4.1 on Data Sheet 4.1.
    c. The food material containing the meal worms should be thoroughly mixed before the second sample is taken.
    d. For the second sample sift through the food and collect meal worms.
    e. Record the total number captured in the second sample and the number of marked individuals in the second sample on Table 4.1 on Data Sheet 4.1.
2. Your instructor can capture, mark, and release animals prior to the lab exercise. You will perform the second sampling.
    a. Your instructor will identify the animal being sampled, the method being used to capture the animals, and how they are marked.
    b. Your instructor will assign responsibilities for collecting the data.
    c. Your instructor will tell you how many individuals were marked and released. Record this number on Table 4.1 on Data Sheet 4.1.
    d. You will conduct the second sampling and record the total number of individuals captured in the second sample and the number of marked individuals in the second sample on Table 4.1 on Data Sheet 4.1.
3. You will perform both sampling activities.
    a. Your instructor will identify the animal being sampled, the method being used to capture the animals, and how they are marked.
    b. On day one you will capture, mark, and release the animals being sampled and record the number of marked individuals released on Table 4.1 on Data Sheet 4.1.
    c. On day two you will resample the population and record the total number of individuals captured in the second sample and the number of marked individuals in the second sample on Table 4.1 on Data Sheet 4.1.

EXERCISE 4
ESTIMATING POPULATION SIZE

Name:_____

Section:_____

Date:_____

Data Sheet 4.1

Table 4.1 Mark-Recapture Estimate of Population Size

Number of marked individuals released initially	Total number of individuals captured in the second sample	Number of marked individuals captured in the second sample

Analysis

Use the data from Table 4.1 in the following formula to determine an estimate of the size of the population of your animal.

$$\text{Total Population} = \frac{\text{Total Individuals in 2}^{nd}\text{ Sample}}{\text{Number of Marked Individuals in 2}^{nd}\text{ Sample}} \times \text{Number Marked Initially}$$

Total Population Size	

Questions

1. List problems you encountered that would reduce the accuracy of your estimate.

   a.

   b.

   c.

EXERCISE 4
ESTIMATING POPULATION SIZE

Data Sheet 4.2

2. Would each of the following factors cause you to have a high, low, or unchanged estimate of the size of the population? Explain each response.

    a. Stress on the initially captured animals affected their survival.

    b. Cold weather reduced the movement of animals.

    c. The species you chose to sample was in the process of migrating north.

    d. The time between the first and second samples was 3 months.

    e. Sampling took place near a water hole used by the species of animal.

EXERCISE 5
POPULATION DYNAMICS

Purpose & Objectives
In this exercise we will use a species from the duckweed family (*Lemna minor*) to explore population dynamics. Population dynamics include investigating factors that affect population size, composition, and distribution. After completing this exercise the student will be able to:
1. Connect major concepts of population dynamics to human populations and the impact it has on the environment.
2. Understand the concepts of exponential growth and logistic growth in populations.
3. Examine density-dependent population growth.
4. Gain an awareness of the extent to which population growth impacts resources.

Introduction
One of the basic units of study in ecology is population dynamics. A population is a group of organisms of the same species occupying the same area at the same time. The study of population dynamics focuses on studying the size, age distribution, and biological and environmental factors that affect the individuals of a population. The size of a population changes over time as the growth rate is affected by the number of births and deaths as well as individuals who leave the population (emigration) and individuals who move into the population (immigration). This balance between the number of individuals entering a population (births and immigration) and the number leaving the population (deaths and emigration) determines whether a population is growing or declining. Populations typically have an innate tendency to grow in size over time. A species' inherent reproductive capacity, or ability to produce offspring, is known as its biotic potential (r).

Exponential and Logistic Population Growth
A population in ideal environmental conditions experiences little resistance against its growth. These conditions lead to what is called an intrinsic rate of growth, which will result in a population that will increase as fast as it possibly can. In this situation, the birth rate far exceeds the death rate and the population will experience exponential growth. This growth pattern is expressed in the growth curve below.

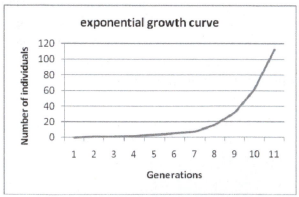

For many organisms, growth is limited by the availability of resources, such as the amount of space, amount of light and availability of nutrients. The availability of these resources will determine the carrying capacity of that particular environment for a specific population. Carrying capacity, K, is the number of individuals of a species that the environment can support over time. Carrying capacity is not a fixed variable and can fluctuate over time as environmental conditions change. As a population increases and approaches its environment's carrying capacity, it will experience environmental resistance against continued growth. Therefore, the population will grow until it reaches the carrying capacity; then it will hover around the carrying capacity, K, of the environment. This growth pattern results in an S-shaped curve and is known as logistic growth as shown below.

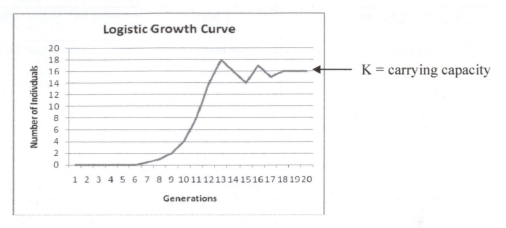

Comparing r-strategists and K-strategists

In general, r-strategists are organisms that typically grow at their maximum rate of biotic potential, r_{max}, and demonstrate exponential growth curves. Some examples of r-strategists include bacteria, many insects, weeds, and some small rodents. However, these organisms will not grow exponentially indefinitely. Inevitably, these organisms will experience a population "crash" as they are regulated by density-independent factors in the environment. Density-independent factors are abiotic factors such as weather conditions and natural events, for example, fires and floods, that dramatically reduce the population size in a short period of time.

K-strategists are organisms that demonstrate a logistic population growth curve. Some examples of K-strategists include but are not limited to birds, predatory fish, trees and larger mammals such as deer, lions, and whales. Initially, when a new population is established, these organisms experience rapid growth as resources are plentiful when the density of the population is low. However, K-strategists' growth is eventually limited as they reach the carrying capacity. The carrying capacity is determined by density-dependent factors that become more effective as the population in that environment (density) increases. Density-dependent factors can be both abiotic (nutrients, dissolved oxygen in aquatic ecosystems, space) and biotic (disease, predation, limited prey).

Table 5.1 Life History Characteristics of General K- and r-Strategists

Characteristics	K-Strategist	r-Strategist
Environmental stability	Stable	Unstable
Size of organism	Large	Small
Length of life	Long, most live to reproduce	Short, most die before reproducing
Number of offspring	Small number produced, parental care provided	Large number produced, no parental care
Primary limiting factors	Disease, Predation, Food Supply, Nutrient Availability (abiotic and biotic)	Weather, Fires, Floods (abiotic)
Population growth pattern	Initially exponential followed by a stable equilibrium stage near carrying capacity	Exponential growth followed by a population crash

Table 5.1 summarizes the characteristics typical of K- and r-strategists. In reality, many organisms do not clearly fit into one category or the other. However, the classification in Table 5.1 is still useful for ecologists when analyzing life history characteristics of organisms.

In this investigation you will examine the growth of a common aquatic plant, duckweed (*Lemna minor*). Duckweed is a free-floating freshwater plant that is common in many ponds and lakes. Populations of duckweed often cover the surface of these freshwater habitats. Typically, a duckweed plant produces one to three elliptical, leaf-like structures known as fronds that float on the surface and a root-like structure that hangs vertically in the water column (see image below). Although duckweed is a flowering plant and can reproduce sexually, its primary method of reproduction is asexual reproduction in which a parent plant splits into two individuals as new fronds are added. By this method of reproduction a population of duckweed can grow rapidly in water with a high nutrient content and rapidly cover the surface of nutrient rich (eutrophic) lakes and ponds.

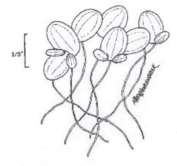

Image from USDA's Plant Database

Materials

Lemna minor

Pond water

Beakers (between 100 ml and 300 ml beakers are appropriate)

Inoculating loops or forceps for transfer

Rulers

Stereo dissecting scopes

Magnifying glass

Fluorescent or natural light source

Calculator

Procedure

1. Examine a single duckweed plant under the stereo dissecting scope and make a sketch of the plant on Data Sheet 5.1. Include the magnification under the sketch.
2. The instructor will divide the class into groups [note: groups should be a minimum of 3 but no more than 4 when possible].
3. Each group is responsible for designing an experimental set-up to test the question, "*What type of growth curve does <u>Lemna</u> <u>minor</u> exhibit?*" You should allow approximately 2 weeks' worth of growth in order to establish a growth curve. Record your experimental design in part B on Data Sheet 5.1. This will include:
    a. Hypothesis – Should include what type of growth curve your group believes *Lemna minor* will exhibit.
    b. Procedure – A description of the step-by-step process your group took to answer the question above. This should include a description of the experimental set-up, control group(s), independent and dependent variable(s) manipulated.
    c. Data Collection – Describe what type of data your group will record and how often it will be collected.
4. In part C on Data Sheet 5.1, you will create a data table that expresses the measurements taken during your experiment. Be sure your table has a title and all columns labeled appropriately.
5. Use the data from the table you created in part C to plot a growth curve of *Lemna minor*.

EXERCISE 5
POPULATION DYNAMICS

Name:_____
Section:_____
Date:_____

Data Sheet 5.1

Part A: Sketch of Lemna minor

Magnification = _____

Part B: Experimental Design

Question: *What type of growth curve does* <u>Lemna</u> <u>minor</u> *exhibit?*

Hypothesis:

_____

_____

Procedure:

_____

_____

_____

_____

_____

_____

_____

Data Collection:

_____

_____

_____

Part C: Data Table

Part D: Growth Curve of a Lemna minor population

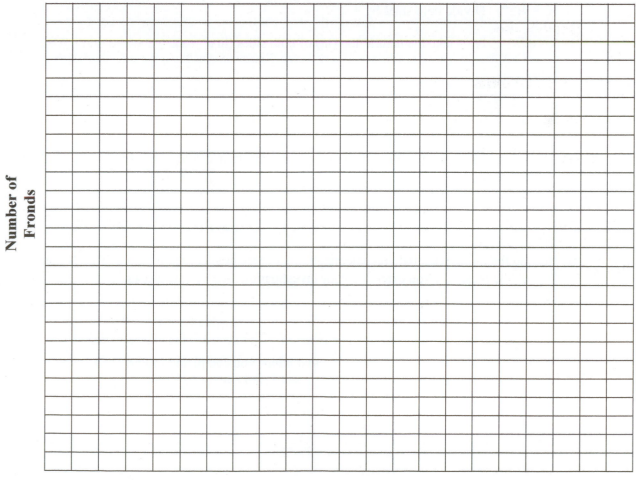

Number of Fronds

Time

EXERCISE 5
POPULATION DYNAMICS

Name:_____
Section:_____
Date:_____

Data Sheet 5.2

Analysis

1. Calculating biotic potential of *Lemna minor*:

   Biotic potential for an organism can be calculated using the following equation: $N_t = N_0e^{rt}$ to calculate the biotic potential of duckweed.
   N_t = the number of individuals at time t
   t = how long you ran the experiment in days
   N_0 = the number of individuals at time 0
   e = the natural logarithm
   r = biotic potential (intrinsic rate of increase)
   For example, suppose there were 4 fronds on day 0 and 12 fronds after one week. What is the population's intrinsic rate of increase over this time period? Use the formula $N_t = N_0e^{rt}$, where N_t = 12, N_0 = 4, and t = 7 days.

   $12 = 4(e^{4r})$ Divide both sides by 4.

   $3 = e^{4r}$ Now take the natural log (ln) of both sides (ln e^x = x).

   $1.098 = 4r$

   $r = 0.275$ per day

   (A) Based on the data you collected from your investigation, calculate the biotic potential for *Lemna minor*.

   (B) Using the biotic potential you calculated in part (A), predict the number of fronds that would be present if you extended your investigation time by two weeks.

2. Based on the population growth curve of *Lemna minor* you completed in Part D of Data Sheet 5.1, what type of population growth would you say *Lemna minor* exhibits?

3. Which reproductive strategy (r-strategist or K-strategist) do you think *Lemna minor* most closely matches? Support your answer by using Table 5.1 and providing the life history characteristics of *Lemna minor* that you think they share with this strategy.

4. Duckweed populations can rapidly increase when nutrients are plentiful. Describe human activities that could create a situation in a lake or pond where nutrients for duckweed would be unlimited. Predict what changes could occur in those freshwater ecosystems if the duckweed's population growth were exponential over a long period of time.

Exercise 6
Historical Changes in Human Population Characteristics

Purpose & Objectives

It is possible to study changes in many human population characteristics by gathering information from cemetery records and obituaries, since cemetery records provide information about the past while obituaries provide information about current times. After completing this exercise, the student will be able to:
1. Describe changes in human mortality and survivorship between past and modern times.
2. State how changes in human mortality and survivorship have influenced population growth.
3. Describe social, biological, and economic factors that contribute to human survivorship.

Introduction

The survival rate of humans in North America has increased significantly in the past 100–200 years. In 1850, life expectancy at birth in the United States was less than 40 years. Today, in the United States, life expectancy at birth is approximately 78 years. This same pattern of increased life expectancy has occurred in the other developed countries of the world. Even less-developed countries have shown significant increases. Improved nutrition, vaccines, antibiotics, preventive medicine, lifestyle changes, control of workplace hazards, and new technologies are a few of the reasons for improved life expectancy. Of particular note is the decline in infant and youth mortality in North America during the past 100 years.

Increasing life expectancies have an impact on population growth rates. The growth of the human population is primarily determined by the birth rate and death rate. As people live longer, they contribute to the population for a longer period. So even with falling birth rates the population can continue to grow.

As the population ages, there are several important demographic and economic considerations. For example, there will be a smaller proportion of the population in the workforce, there will be an increased cost of government social programs for retired persons, and older people will need specialized health care.

One of the ways to visualize how a change in the death rate affects the population is to construct a graph of the number of people who survive to each age. This is known as a survival curve. In this exercise we will construct four survival curves that will allow us to look at differences between men and women, and between current individuals and those that lived over a century ago.

The class will collect four sets of data that show age at death of individuals who lived in your local region. The first two sets of data will be for men and women who died before 1900. The second two sets of data will be for men and women who died within the past five years. Once these data have been collected, the number of individuals surviving to each age can be calculated and a survival curve can be constructed.

Procedure

1. The class goal is to collect information on the *age at death* for 100 males and 100 females who died before 1900 and for 100 males and 100 females who died within the last five years. (If that is not possible, a smaller data set can be collected and converted to percentages.) You will need to rely on names and other information provided to determine whether a person is a male or female.

   It is also important to make sure that the same individual is not counted more than once. Therefore, to divide the work load and prevent duplication, your instructor will provide each student with specific directions on which resources to use and what data to collect. Write that information on Table 6.1.

Table 6.1 Individual Student Responsibilities

	Name of Source to Search for Data (Cemetery Records or Obituary Records)	Number of Records to Obtain	In This Column, Record the *Age at Death* for Your Assigned Population. (For Example 19, 37, 99, 101, 6, 82, etc.)
Pre-1900 males			
Pre-1900 females			
Current males			
Current females			

2. Visit local cemeteries, or obtain cemetery information online for individuals who died prior to 1900. Obtain obituary information from local newspapers or online sources for individuals who have died within the past five years. Record your information on Table 6.1. Many cemeteries have collected all the information from tombstones and provide the information in their office, at local libraries, and online. To find online resources for local cemeteries simply type in the name of your city or county and the phrase "cemetery records." Obituary information is available through local newspapers, which also have the information online.

3. In class, combine the data from all students and enter the data on the number of individuals who died at each age on Table 6.2 Human Survival Data on Data Sheet 6.1.

4. Calculate the number of individuals surviving to each age by the method shown below and record this information on Table 6.2 Human Survival Data on Data Sheet 6.1.

Age at death (Years)	Number that died	Percent surviving
0	0	100 − 0 = *100*
0—0.99	10	100 − 10 = *90*
1—4.99	15	90 − 15 = *75*
5—9.99	12	75 − 12 = *63*

5. Use different colored pencils to plot the four sets of survival data (males before 1900, females before 1900, males current, and females current) on Graph 6.1 Survival Curve. Alternatively enter the data from the 5 shaded columns of Table 6.2 Human Survival Data Sheet into a data base like Excel and use the scatter graph (X-Y graph) option to create the graph.

6. Analyze the data and the reasons for change by answering the questions on Data Sheet 6.3.

EXERCISE 6
HISTORICAL CHANGES IN HUMAN
POPULATION CHARACTERISTICS

Name:_____

Section:_____

Date:_____

Data Sheet 6.1

Table 6.2 Human Survival Data

Age at death (years)	Pre-1900 Cemetery Data				Current Obituary Data (last 5 years)			
	Male		Female		Male		Female	
	Number that died	Percent surviving	Number that died	Percent surviving	Number that died	Percent surviving	Number that died	Percent surviving
Live birth 0	0	100	0	100	0	100	0	100
0–0.99								
1–4.99								
5–9.99								
10–14.99								
15–19.99								
20–24.99								
25–29.99								
30–34.99								
35–39.99								
40–44.99								
45–49.99								
50–54.99								
55–59.99								
60–64.99								
65–69.99								
70–74.99								
75–79.99								
80–84.99								
85–89.99								
90–94.99								
95–99.99								
Over 100								

EXERCISE 6
HISTORICAL CHANGES IN HUMAN
POPULATION CHARACTERISTICS

Name:_____
Section:_____
Date:_____

Data Sheet 6.2

Graph 6.1 Survival Curve

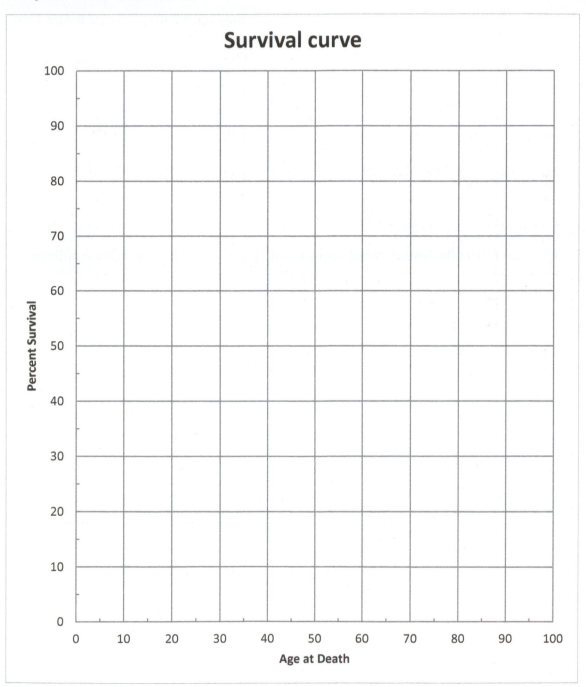

EXERCISE 6
HISTORICAL CHANGES IN HUMAN
POPULATION CHARACTERISTICS

Name:_____

Section:_____

Date:_____

Data Sheet 6.3

Analysis

1. Examine Graph 6.1.
   a. List two differences in the patterns of survival of men and women who died before 1900.

   b. What factors (biological, social, economic) do you think contributed to these differences?

2. Examine Graph 6.1.
   a. List two differences in the patterns of survival of men and women who died within the last 5 years.

   b. What factors (biological, social, economic) do you think contributed to these differences?

3. Examine Graph 6.1.
   a. List two differences in the patterns of survival of men who died prior to 1900 with those who died within the last 5 years.

   b. What factors (biological, social, economic) do you think contributed to these differences?

4. Examine Graph 6.1.
   a. List two differences in the patterns of survival of women who died prior to 1900 with those who died within the last 5 years.

   b. What factors (biological, social, economic) do you think contributed to these differences?

EXERCISE 7
HUMAN POPULATION DYNAMICS

Purpose & Objectives

Human population dynamics addresses the growth of global human populations now and in the future. In this exercise, you will be using US Census Bureau data to predict what the population of three different countries populations will be in 25 years. For each country you will need to fill in the data tables provided and use the formulas provided to determine final predicted population. After completing this exercise, the student should able to:

1. Calculate doubling time and growth rates for developed and developing countries, using the population growth formula.
2. Identify and describe the factors that contribute to growing human populations.
3. Discuss the effects of projected decreases in growth rates over time.
4. Describe the impact of human population growth on consumption rates of global resources and connect how this may affect the sustainability of human societies and quality of life.

Introduction

The study of human population growth is important to both ecologists and environmental scientists as it impacts the health and function of the Earth's natural systems. Scientists use demographic, social, and economic indicators to predict human population growth over time. The impact that humans have on the planet is often determined by population size, available resources in a region, and the stage of social development within that population.

The growth rate of a species is the difference between the number of births and number of deaths in any given population. Human population growth can be calculated given the growth rate, initial population, and time interval by using a simple mathematical formula, $p_f = p_i * e^{rt}$.

p_f = final population
p_i = initial population
e = a constant with the value of 2.7183
r = rate of growth as a decimal. For example 1.3% = 0.013
t = time in years.

If we know the initial population and growth rate, we can use this mathematical relationship to predict what the population will be after a given time period (t). For example, the world's population in 2010 was 6.86 billion people. By using the world's growth rate in 2010 (see Table 10.1) we can predict what the global population will be in 50 years (*Tip: it is often easier to calculate the exponent (0.011 x 50 = 0.55) and then use the e^x key to yield 1.733 when solving these types of equations*):

$$P_f = 6.86 \times 10^9 * e^{(0.011)(50)} = 11.89 \text{ billion (in the year 2062)}$$

We can also use the "rule of 70" to calculate the doubling time (dt = 70/r) for any region or country. Assuming a constant growth rate for the world at 1.1%, we could predict that the world population will double by the year 2074.

$$dt = 70/r \quad = \quad 70/1.1 = 64 \text{ years} \quad = \quad 2010 + 64 \text{ years} = \text{the year 2074}$$

Countries throughout the world have different growth rates, which will affect their population size, doubling time, and impact upon the environment. There are many demographic indicators, such as fertility and growth rates, that allow us to make general predictions about human population growth in any one particular region or area. Generally speaking, more developed countries tend to have much lower growth rates than less developed countries. Today, the overall growth rate (r) for more developed nations is 0.3 percent and for less developed nations 1.3 percent. Therefore the time it would take for a more developed nation's population to double in size is 233 years and for less developed nations is 54 years. At the current population growth rates (2010), in one year the populations of more developed nations increased by 4.9 million people. During this same period, however, less developed nations increased their population by 73.5 million people. The human population reached 7 billion people in 2012.

Table 7.1 Comparison of World and Regional Population Growth from 1998 to 2010

Region	YEAR: 1998		YEAR: 2010	
	Growth Rate	Doubling Time	Growth Rate	Doubling Time
World	1.4	50	1.1	64
More Developed Countries	0.4	175	0.3	233
Less Developed Countries	1.8	39	1.3	54
Africa	2.5	28	2.27	31
Asia	1.6	44	1	70
North America	1	70	0.9	78
Latin America	1.7	41	1.1	64
Europe	0.2	350	0.1	700
Russia	0.3	233	-0.1	n/a
Oceania	1.5	47	1.5	47

Sources: United Nations' World Population Prospects, 1998 and 2010 Revisions.

Population growth is regulated by four main demographic factors: (1) birthrates, (2) death rates, (3) immigration, and (4) emigration. Large emigration (outward migration) reduces the population growth rate and large immigration (in-ward migration) increases the growth rate of the receiving country. Demographers often classify countries according to their stage of demographic transition. The demographic transition model generalizes changes in human population over time as the society experiences socioeconomic progress. However, this is an idealized, Eurocentric model of progress and it is not necessarily indicative of how human populations in every country or region will change over time.

Figure 7.1 Demographic Transition Model

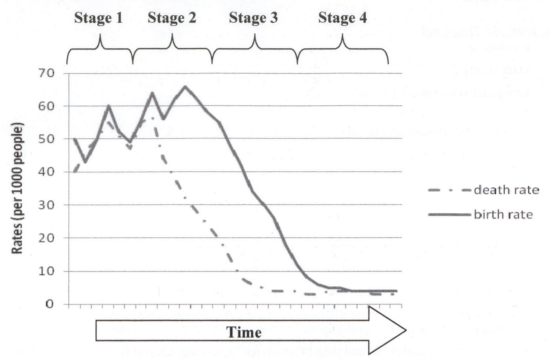

Stage 1 – high birth rates and high death rates = low growth rate

Stage 2 – high birth rates and decreasing death rates = high growth rate

Stage 3 – decreasing birth rates and low death rates = low growth rate

Stage 4 – stable, low birth rates and death rates = low to zero growth rate

In order to bring the world's rapidly growing human population under control, we need to understand and manage the factors that control both birth and death rates. To reach zero population growth (ZPG), the crude birthrate has to decrease faster than the crude death rate until the growth rate approaches zero percent. Although birth rates are dropping in many countries, the overall number of children added to the population each year is enormous. That is, even though many women are having fewer babies, there are millions of women having babies each year. Can countries achieve zero population growth? Yes they can, but it will take knowledge, commitment, and access to resources. Countries will take different amounts of time to achieve ZPG, depending on the growth rate, access to education (especially for women), and birth control.

Procedure

Materials Needed

Calculator

Data sheets

Computer with Internet access

1. Go to the Population Reference Bureau website located at:

   http://www.prb.org

2. Click on the DataFinder tab.

3. Click on the United States & International Profiles tab.

4. Next, select "**International**" and then "**Countries**".

5. Select your first country (only one) from the list below.

   a. Choose from countries: United States, Belgium, France, Sweden, China

6. Now you should see data about your first selected country.

7. We want to compare three different countries at one time. Therefore, click on the "**change locations/indicators**" next to your country's name at the top of the page.

8. Select "**International**" and then "**Countries**" again and choose two more countries—one from list *a* and one from list *b* below:

   a. For Country 2 select from: Ukraine, Japan, Bulgaria, Latvia

   b. For Country 3 select from: Congo (Brazzaville), Haiti, Iraq, Ethiopia, Dominican Republic, Guatemala

9. Next, to the right of the **Locations** column you will find the **Indicators** column. Change the indicators that are currently selected for data. Select only the "**demographics, economics, education,** and **environment**" from this list; then click "**continue.**"

10. The demographic indicators should be the first results you see. Fill in data in Tables 7.2 and 7.3 for demographic, economic, and social indicators.

11. Record the environmental impact for each of your countries in Table 7.4.

EXERCISE 7
HUMAN POPULATION DYNAMICS

Name:_____

Section:_____

Date:_____

Data Sheet 7.1

Table 7.2 Demographic Indicators

2022	Country 1: United States	Country 2: Ukraine	Country 3: Dominican Rep.
Population Mid-Year	332.8	41.0	11.2
Population Growth Rate	0.4%	−0.6%	1.0%
Rate of Natural Increase	0.1%	−1.1%	1.2%
Population Projection – mid-2050	375.4	33.9	13.2
Total Fertility Rate	1.7	1.0	2.3
Population < 15 Years Old	18%	15%	27%

Table 7.3 Social and Economic Indicators

2022	Country 1: United States		Country 2: Ukraine		Country 3: Dominican Rep.	
GNI per Capita (US$)	$70.480		$13.860		$19.730	
GDP (million PPP$)	20,94 Trillion		155.6 Billion		78.84 Billion	
Population Living Below $2/day (US$)	2.6%		1.3%		15.9%	
Per Capita Expenditure on Health Care (US$)	$11.945		$248		$491	
Secondary School Enrollment by Gender (net)	Males:	Females:	Males:	Females:	Males:	Females:

EXERCISE 7
HUMAN POPULATION DYNAMICS

Name:_____

Section:_____

Date:_____

Data Sheet 7.2

Table 7.4 Environmental Impact

2022	Country 1: United States	Country 2: Ukraine	Country 3: Dominican Rep.
CO_2 Emissions per Capita (metric tons)	15.24	4.15	2.36
Motor Vehicles per 1000	816.4	184	442
Natural Habitat Remaining	30%	54%	65%
Population Using Improved Sanitation	96%	69%	54.4%

Analysis

1. Use the formula $pf = pi * e^{rt}$ and the appropriate demographic indicators you recorded in Table 7.2, to calculate what the population for each of your three countries would be in 2050. Compare your calculations of the mid-2050 population to the population projections provided by the Population Reference Bureau that you recorded in Table 7.2. Discuss what factors or assumptions could account for any differences you see between the two predictions.

2. Look at the data you recorded in Table 7.2. When you compare the three countries, how are the age <15, the total fertility rate, and the rate of natural increase related to the total population growth rate?

EXERCISE 7
HUMAN POPULATION DYNAMICS

Name:_____
Section:_____
Date:_____

Data Sheet 7.3

3. Based on the social and economic data you recorded in Table 7.3 and the environmental impacts you recorded in Table 7.4, fill in the chart below with the social, economic, and environmental characteristics of a rapidly growing population versus a declining population.

Indicators:	Rapidly Growing Population	Declining Population
Social Factors	1. High birth rates 2. Improved Healthcare 3. Immigration	1. Aging population 2. Low birth rates 3. Emigration
Economic Factors	1. Increased Labor Force 2. Market Expansion 3. Demographic Dividend	1. Labor Shortages 2. Increased Dependancy Ratio 3. Reduced Consumer Base
Environmental Factors	1. Resource consumption 2. Habitat loss and land Conversion 3. Pressure on Ecosystem	1. Land Reclamation 2. Ecological Restoration 3. Decreased Habitat Conversion

4. For your fastest growing country, calculate how long it will take for the population to double in size. Describe some strategies that would be effective for slowing this population's growth.

EXERCISE 7
HUMAN POPULATION DYNAMICS

Name:_____

Section:_____

Date:_____

Data Sheet 7.4

5. Country X, whose current population is 12.5 million people, is experiencing a growth rate of 1.4% per year. Country Y, whose current population is 8 million people, is experiencing a growth rate of 2.8% per year. Assuming the growth rate for the two countries stays the same, answer the following questions.
    a) Calculate how many years it would take for Country X to double in population size.
    b) Calculate how many years it would take for Country Y to double in population size.
    c) In 50 years, what will be the population of both countries?

6. Examine the population pyramid for the two countries below. For each country, identify what stage of demographic transition you would place them in and justify your answers.

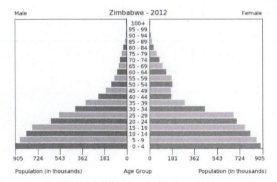

Data: US Census Bureau, 2012

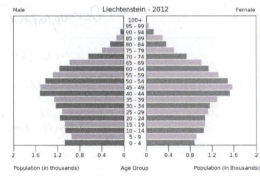

Data: US Census Bureau, 2012

EXERCISE 8
PLATE TECTONICS

Purpose & Objectives

This exercise will utilize information from the United States Geological Survey (USGS) to allow students to gain an understanding of plate boundaries and their movement. After completing this exercise, the student will be able to:

1. Compare the strength of earthquakes by using the Richter magnitude scale.
2. Describe different kinds of plate boundaries and their movement patterns.
3. Explain how the movement of tectonic plates leads to earthquake activity and the formation of volcanoes, trenches, ridges, and mountain ranges.

Introduction

The upper layer of the Earth is a relatively rigid layer known as the lithosphere. The lithosphere is broken up into a large number of plates known as tectonic plates that lie on top of a zone of the mantle known as the asthenosphere, which contains semi-solid material that allows for the movement of tectonic plates over its surface over long periods of time. (See Figure 8.1.) Plate tectonic theory involves concepts associated with the movement of tectonic plates and the interactions that occur at plate boundaries. Tectonic plates are called oceanic plates if they are located under the ocean and continental plates if they are part of a continent.

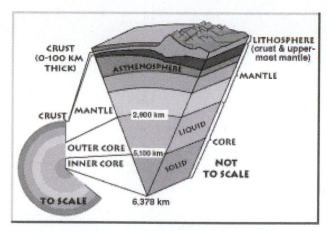

Figure 8.1 Cross Section of the Earth
Source: USGS: *Dynamic Planet*

There are three types of plate boundaries: divergent, convergent, and transform-fault.

- Divergent boundaries exist where plates are moving away from one another. As the plates move away from one another, magma will be forced up through the mantle to fill the void, thereby creating new crust. This type of movement is often referred to as "sea-floor spreading" and forms volcanoes and oceanic ridges, such as the Mid-Atlantic oceanic ridge. As terrestrial plates diverge from one another they form a central valley known as rift. Examples of rift valleys include the East African rift in Kenya and the Rio Grande rift in New Mexico.
- Convergent boundaries exist where plates are slowly moving toward one another. Typically one plate will slide underneath the other and this is known as subduction. Subduction zones are characterized by trench formations and volcanism. Convergence between two oceanic plates (oceanic-oceanic convergence) is responsible for deep trench structures, such as the Mariana trench, which is the deepest point on Earth (11 kilometers or 7 miles deep). Convergence between

an oceanic and continental plate (oceanic-continental convergence) can result in the uplift of the continental plate that forms mountain ranges. The Andes mountain range, along the western coast of South America, formed as the oceanic Nazca plate converged with the continental South American plate. Earthquake activity is common as continental plate uplift occurs. The convergence of two continental plates (continental-continental convergence) also causes continental uplift and produces mountain ranges. The Himalayas were formed as the Eurasian plate converged with the Indian plate.

- Transform-fault boundaries exist where plates are sliding horizontally past one another. This type of movement does not result in construction of crust (as in divergent movement) or in destruction of crust (as in convergent movement). Transform-fault boundaries are characterized by shallow earthquake activity. One of the most famous examples of this type of plate boundary is the San Andreas Fault in California.

Most of the world's volcanoes and earthquake activity occurs in the Circum-Pacific belt commonly referred to as the "Ring of fire." (See Figure 8.2.)

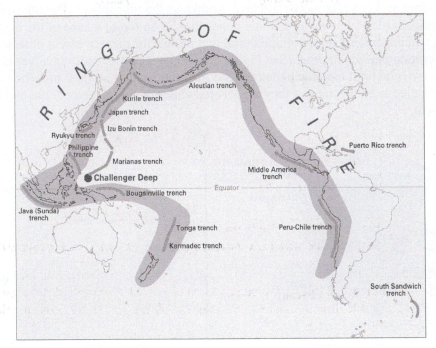

Figure 8.2: Ring of Fire
Source: USGS: *Dynamic Planet*

In order to compare the size of earthquakes to one another, the Richter magnitude scale is used. This scale measures the logarithm of the amplitude of seismic waves as recorded by a seismograph. Worldwide, it is not uncommon for 6.0 magnitude earthquakes to occur several times a month. Since the Richter scale is logarithmic an earthquake measuring 8.0 on the Richter scale will have a shaking amplitude ten times larger than a magnitude 7 earthquake and 100 times larger than that of a 6.0 earthquake. However, magnitude cannot indicate potential for property damage or loss of human life since that is also dependent on the density of the population in the region impacted by the earthquake as well as structural quality of commercial buildings and homes in that area.

Procedure

Materials Needed

Colored pencils

Data sheets

Computer with Internet access

Activity

We will use a series of mapping activities to develop an understanding of the role plate tectonics play in the production of earthquakes and volcanic activity. We can also assess the level of hazard to people and property by knowing where active earthquake zones are located. We will be using latitude and longitude to locate earthquakes on the map (see Figure 8.3).

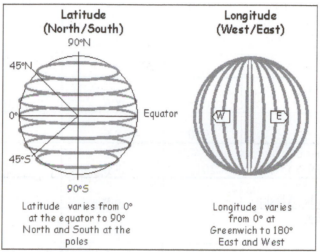

Figure 8.3: Longitude and Latitude
Source: FedStats: Statistical Programs of the U.S. Government

1. Table 8.1 lists the largest earthquakes, in magnitude, in the world since 1900. Plot the location of each earthquake on the world map provided on Data Sheet 8.1. For each plot record the number (1-17) and the magnitude directly underneath. [Note: You may want to choose a colored pencil to record your plots so they are easily seen on the map.]

2. Use the USGS website to find the major plates of the earth (you can go to USGS.gov and search for "plates of the earth"). Draw the major plates of the Earth onto to your map. Choose different colors to represent divergent, convergent, and transform-fault plate boundaries.

3. Use a global population density map to identify areas that could experience high death rates and economic damage from earthquakes based on areas or regions with high population density that also lie near active fault zones. [Note: You can access global density maps at http://sedac.ciesin.columbia.edu/data/collection/gpw-v3.]

4. Use information from your textbook or the Internet to map the following geological structures on your map: mid-Atlantic ridge, Aleutian trench, Mariana trench, Himalayan mountain range, Andes mountain range.

5. Use the USGS Volcano Hazard program found at http://volcanoes.usgs.gov/ to identify the major locations for United States volcanoes. Plot any U.S. volcanoes that are listed as advisory, watch, or warning status on your map.

Table 8.1 Largest Earthquakes in the World Since 1900

	Location	Year	Magnitude	Latitude	Longitude
1	Chile	1960	9.5	38.29 S	73.05 W
2	Prince William Sound, Alaska	1964	9.2	61.02 N	147.65 W
3	Off the West Coast of Northern Sumatra	2004	9.1	3.3 N	95.78 E
4	Near the East Coast of Honshu, Japan	2011	9	38.322 N	142.369 E
5	Kamchatka	1952	9	52.76 N	160.06 E
6	Offshore Maule, Chile	2010	8.8	35.846 S	72.719 W
7	Off the Coast of Ecuador	1906	8.8	1 N	81.5 W
8	Rat Islands, Alaska	1965	8.7	51.21 N	178.5 E
9	Northern Sumatra, Indonesia	2005	8.6	2.08 N	97.01 E
10	Assam - Tibet	1950	8.6	28.5 N	96.5 E
11	Off the west coast of northern Sumatra	2012	8.6	2.311 N	93.063 E
12	Andreanof Islands, Alaska	1957	8.6	51.56 N	175.39 W
13	Southern Sumatra, Indonesia	2007	8.5	4.438 S	101.367 E
14	Banda Sea, Indonesia	1938	8.5	5.05 S	131.62 E
15	Kamchatka	1923	8.5	54 N	161 E
16	Chile-Argentina Border	1922	8.5	28.55 S	70.5 W
17	Kuril Islands	1963	8.5	44.9 N	149.6 E

Data From: USGS Earthquake Hazard Program (April, 2012)

EXERCISE 8
PLATE TECTONICS

Name:_____
Section:_____
Date:_____

Data Sheet 8.1

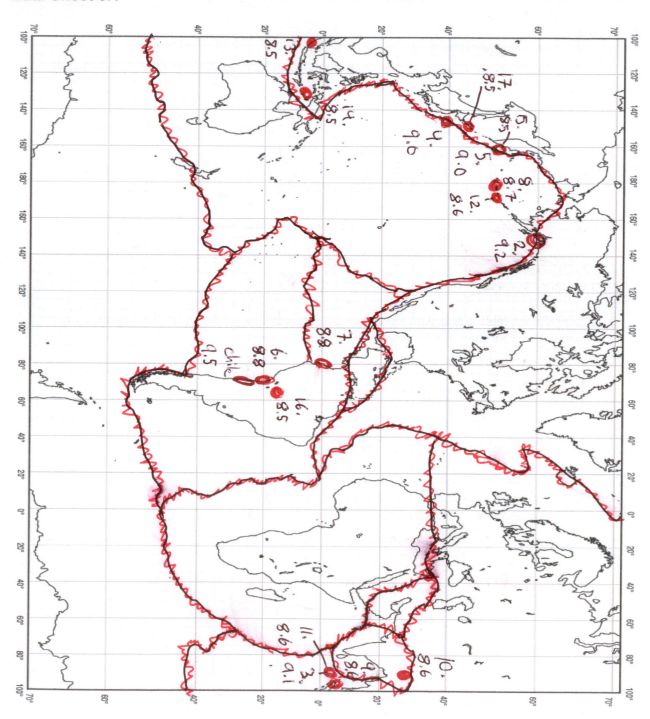

EXERCISE 8
PLATE TECTONICS

Name:_____
Section:_____
Date:_____

Data Sheet 8.2

Analysis

1. Examine your plots of the earthquakes on the map.
    a. Determine if they lie on major plate boundaries. If so, what is the most common plate boundary for the earthquakes you plotted?

    b. Explain how this type of plate movement causes earthquakes.

2. Complete the table below with the appropriate information for the geologic structures you included on your map. See the example provided for you as guidance.

Geologic Structure	Plate Boundary	Plate Movement
Example: Aleutian Trench	*Pacific Plate → North American Plate*	*Convergent / Subduction*
Mariana Trench		
Andes Mountain Range		
Himalayan Mountain Range		
Mid-Oceanic Ridge		

EXERCISE 8
PLATE TECTONICS

Name:_____
Section:_____
Date:_____

Data Sheet 8.3

3. The earthquake recorded in the region of Haiti on January 12th, 2010, had a magnitude of 7.0. Although it does not make the list of the "largest" earthquakes, as seen in Table 8.1, it is ranked number one on the list of the most deadly earthquakes as it took the lives of over 300,000 people (as of 2012 USGS data). Many of the earthquakes in Table 8.1 would not make the list of the top deadliest earthquakes.

   a. How many times larger is the shaking amplitude of the Haiti earthquake in comparison to the one responsible for the nuclear power plant accident in Fukushima, Japan, in 2011 (#4 in Table 8.1)?

   b. Discuss three factors that could cause some low magnitude earthquakes to cause more economic damage and greater loss of life than earthquakes that were of higher magnitude. Give an example of an area or region that is at risk based on what you found in your investigation (#3 in the Procedures).

4. Explain what type of plate boundaries and movement can lead to the formation of volcanoes.

EXERCISE 9
SOIL CHARACTERISTICS AND PLANT GROWTH

Purpose & Objectives

In this exercise we will look at how the size of soil particles and the availability of nutrients affect the growth of plants. After completing this exercise the student will be able to:

1. Determine the water-holding capacity of different kinds of soils.
2. Observe differences in the growth of radish plant roots in different kinds of soil.
3. Measure biomass production of radish plants following two weeks of growth in different soils.
4. Determine the effects of fertilizer on the growth of radish plants.

Introduction

Soil is a complex mixture of mineral materials of different sizes and chemical composition, air, water, decomposing organic matter, and living organisms. The mineral particles of soil are classified based on particle size. See Table 9.1.

Table 9.1 Size of Soil Particles

Kind of Particle	Size Range
Gravel	Greater than 2.0 mm
Sand	Between 0.05 and 2 mm
Silt	Between 0.002 and 0.05 mm
Clay	Less than 0.002 mm

Particle size is extremely important. Soils with many large particles tend to have many pore spaces between them that allow water and air to enter the soil. Soils consisting of many tiny particles have fewer pore spaces and, therefore, water and air do not enter these soils as readily. The infiltration rate (how fast water can enter a soil) and water-holding capacity are greatly influenced by particle size. Porous soils, with large particles, such as sands, allow water to enter readily but do not retain it for very long and dry out quickly. Clay soils with many tiny particles do not allow water to infiltrate quickly, but once they are wet they take a long time to dry out.

Both water and air are important to the growth of roots. Roots carry on aerobic respiration and thus need oxygen to grow. However, roots need to absorb water from the soil that will be used to transport materials through the plant and as a raw material for the process of photosynthesis. Thus, a good soil must have a balance between its ability to hold water and provide oxygen. The soil textural triangle is a commonly used method for describing different kinds of soils based on the percentage of sand, silt, and clay present in the soil. Those near the center of the triangle (various kinds of loams) tend to have a mixture of particle sizes that allows for adequate water-holding capacity but also allows air to penetrate the soil. See Figure 9.1.

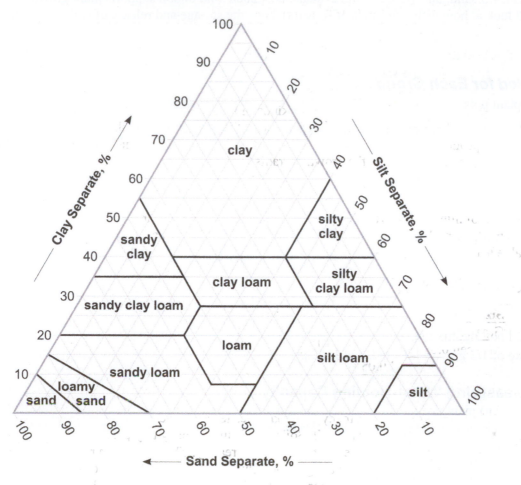

Figure 9.1 Soil Textural Triangle (Courtesy of the USDA.)

The three primary mineral nutrients often in short supply in soil are nitrogen, phosphorus, and potassium compounds. Because of their major roles in the growth and health of plants, they are often referred to as **macronutrients.** Nitrogen is a necessary atom in the structure of proteins and, therefore, is essential for growth. Phosphorus is a component of DNA and RNA, is involved in energy transfer molecules (ATP and other molecules), is involved in photosynthesis and respiration, and is a component of cell membranes. It also is important for the growth of roots and the development of flowers and fruit. Potassium is involved in a wide variety of important metabolic activities, including photosynthesis and the transport of many kinds of molecules throughout the plant. Fertilizer consists of mixtures of these three macronutrients. The mineral content of fertilizer is written as a series of three numbers such as: 20—20—20; 0—20—20; or 6—12—12. Each number indicates the percent by weight of a nutrient available. The first number in the series is the percent of nitrogen, the second number is the percent of phosphorus, and the third number is the percent of potassium.

In addition to the mineral material in soil, organic matter is extremely important. Decomposing organic matter improves the porosity and water-holding capacity of the soil, improves the infiltration rate, and provides nutrients.

In this exercise we will look at five different soils: pea gravel, sand, clay, commercial potting soil, and a mixture of 1/3 gravel, 1/3 sand, and 1/3 clay; and evaluate how each kind of soil supports plant growth. In addition, we will look at how different kinds of soils influence the storage and release of soil nutrients.

Procedure

Materials Needed for Each Group
Ten 8 cm (3 inch) plant pots
Graduated cylinder
Electronic scales
Forceps
Teasing needles
Ruler that measures in millimeters
Five 150 mL beakers
Solution of 20-20-20 fertilizer
The following soil types
> Pea gravel
> Pure sand
> Pure clay
> Commercial potting mixture
> Soil mixture of 1/3 gravel, 1/3 sand, and 1/3 clay

Exercise 9.1: Measuring Water-Holding Capacity
Obtain five 8 cm (3 inch) plant pots.
1. Place pottery shards or other objects over the holes in the pots so that soil does not run out through the holes. This can be somewhat tricky with fine soils, since we want water, but not the soil to be able to pass through the holes.
2. Fill each pot to within 1 cm of the top with one of the five soil mixtures available. Gently press down the soil so that any large air spaces are eliminated. This is particularly important for clay soil.
3. Slowly pour 100 mL of water onto the surface of the soil in each pot and catch and measure the amount of water that passes through the soil and drips from the pot. You may need to pour very slowly for some of the soils. Record the amount of water exiting the pot.
4. If any pots have large amounts of soil washed through the holes in the bottom of the pot when water is pour onto the soil, redo the setup and more effectively cover the holes in the bottom of the pot.
5. Determine the amount of water held in the soil by subtracting the number of milliliters of water exiting the pot from 100 mL and record the data on Table 9.2 on Data Sheet 9.1.

Exercise 9.2: The Effect of Water-Holding Capacity on Growth
1. Use the five pots from Exercise 9.1 in which you measured water-holding capacity.
2. Plant five sprouting radish seeds in each pot 1 cm below the surface.
   The sprouting seeds are easily damaged, so handle them carefully.
3. Add 10 mL of water to each pot each day for two weeks.
4. At the end of two weeks, record the following information on Table 9.3 on Data Sheet 9.1.
   a. Count the number of plants alive in each pot.
   b. Measure the root length of the plants in each pot. In order to do this you will need to carefully remove the plants from the soil.
      i. Gently remove the soil along with the plants from the pot and place it on a tray or several layers of newspaper.

    ii.   Use forceps or teasing needles to carefully pick apart the soil so that you isolate the roots from the soil. You may need to gently spray water on them to get rid of some of the soil particles.

    iii.   Measure the root length of each plant in millimeters, add the root lengths of all plants from each pot together, and enter the total length on Table 9.3 of Data Sheet 9.1.

    iv.   Calculate the average root length of plants by dividing the total root length by the number of plants surviving. Enter the data on Table 9.3 of Data Sheet 9.1.

c.   Measure the total biomass of all the plants in each pot.

    i.   Place the plants from each pot on an electronic balance and determine the total biomass in grams.

    ii.   Record the total biomass on Table 9.3 on Data Sheet 9.1.

    iii.   Calculate the average biomass of plants by dividing the total biomass by the number of plants surviving. Enter the data on Table 9.3 of Data Sheet 9.1.

Exercise 9.3: The effect of soil nutrients on growth

Obtain five 8 cm (3-inch) pots and fill each to within 1 cm of the top with one of the five soil mixtures available.

1.   Slowly pour 100 mL of a commercially available 20-20-20 plant food solution onto the surface of the soil in each pot and catch and discard any fertilizer solution that passes through the soil and drips from the pot. You may need to pour very slowly for some of the soils.

2.   If any pots have substantial amounts of soil washed through the holes in the bottom of the pot when the fertilizer solution is poured onto the soil, redo the setup and more effectively cover the holes in the bottom of the pot.

3.   Plant five sprouting radish seeds in each pot 1 cm below the surface.
The sprouting seeds are easily damaged, so handle them carefully.

4.   Add 10 mL of water (without plant food) to each pot each day for two weeks.

6.   At the end of two weeks, remove the plants from the soil and record the following information about the plants grown in different soils on Table 9.4 on Data Sheet 9.1. (See the directions for Exercise 9.2.)

a.   Count the number of plants alive in each pot.

b.   Measure the root length of the plants in each pot. Add the root lengths of all plants in a pot together and enter the total root length on Table 9.4 of Data Sheet 9.1.

    i.   Calculate the average root length for each pot by dividing the total root length by the number of plants surviving in each pot.

    ii.   Record the data on Table 9.4 on Data Sheet 9.1.

c.   Measure the total biomass of the plants in each pot.

    i.   Calculate the average biomass per pot.

    ii.   Enter the data on Table 9.4 of Data Sheet 9.1.

d.   Compare your data from Table 9.4 with those from Table 9.3 to see if the plant food had any effect on the growth of the radish plants.

EXERCISE 9
SOIL CHARACTERISTICS AND PLANT GROWTH

Name:_____
Section:_____
Date:_____

Data Sheet 9.1

Table 9.2 Water-Holding Capacity

Kind of soil	Milliliters of water held
Pea gravel	
Sand	
Clay	
1/3 gravel, 1/3 sand, 1/3 clay	
Potting soil	

Table 9.3 Effect of Soil Type on Plant Growth

Kind of soil	Number of plants surviving	Root length (mm)		Biomass (grams)	
		Total length	Average length	Total biomass	Average biomass
Pea gravel					
Sand					
Clay					
1/3 gravel, 1/3 sand, 1/3 clay					
Potting soil					

Table 9.4 Effect of Nutrients on Plant Growth

Kind of soil	Number of plants surviving	Root length (mm)		Biomass (grams)	
		Total length	Average length	Total biomass	Average biomass
Pea gravel					
Sand					
Clay					
1/3 gravel, 1/3 sand, 1/3 clay					
Potting soil					

EXERCISE 9
SOIL CHARACTERISTICS AND PLANT GROWTH

Name:_____
Section:_____
Date:_____

Data Sheet 9.2

Analysis

Construct a bar graph of the relationship between the type of soil and the average root length and the relationship between type of soil and average biomass of those plants grown without fertilizer. You may want to put the data into Excel and use the bar graph option to generate the two graphs.

Graph 9.1 Average Root Length

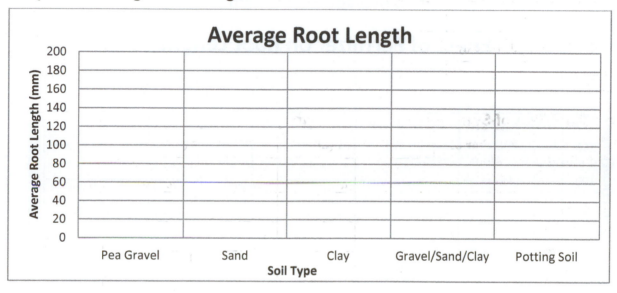

Graph 9.2 Average Plant Biomass

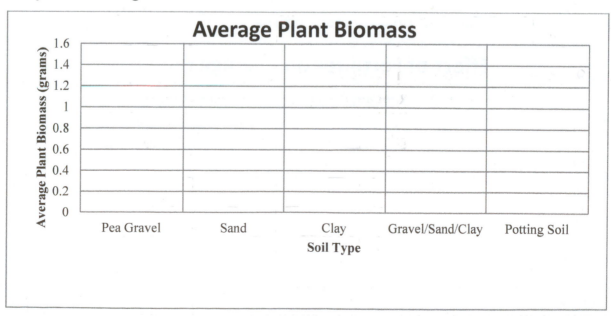

EXERCISE 9
SOIL CHARACTERISTICS AND PLANT GROWTH

Name:_____

Section:_____

Date:_____

Data Sheet 9.3

Construct a bar graph that compares the average root length of plants grown with and without fertilizer for each of the 5 soil types. For each soil type you will have two bars, one for the average root length without fertilizer and one for average root length with fertilizer. Construct a second bar graph that compares average biomass with and without fertilizer. You may want to input the data into Excel and use the bar graph option to generate the graph.

Graph 9.3 Effect of Fertilizer on Average Root Length

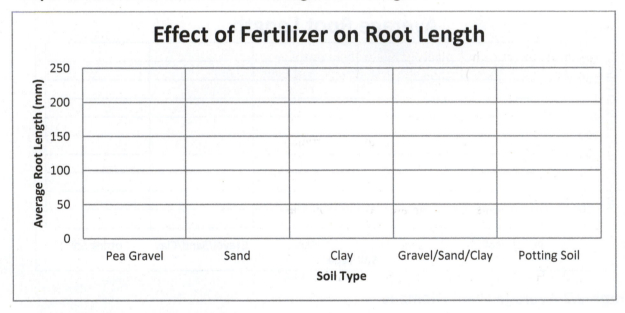

Graph 9.4 Effect of Fertilizer on Average Biomass

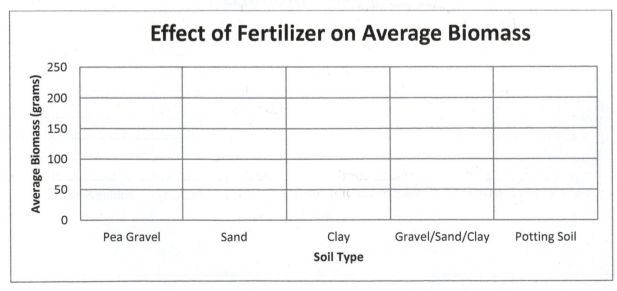

EXERCISE 9
SOIL CHARACTERISTICS AND PLANT GROWTH

Name:_____
Section:_____
Date:_____

Data Sheet 9.4

Questions

1. Compare the water-holding capacity of the soils.
   a. Which soils had the greatest water-holding capacity?

   b. How did particle size affect water-holding capacity?

2. Compare the growth of plants grown without fertilizer.
   a. Which soil(s) had the greatest average root length?

   b. How was root length related to soil particle size?

   c. Which soils had the greatest average biomass?

   d. How was biomass related to soil particle size?

3. Compare the growth of plants that were unfertilized with those that were fertilized.
   a. Did fertilized plants grow better than unfertilized plants in all soil types?

   b. If some unfertilized plants grew as well or better than fertilized plants in some soils, what characteristics of the soil may have contributed to the unexpected result?

4. The commercial potting soil contains considerable amounts of organic matter. How could the organic matter have affected the growth of plants?

EXERCISE 10
STREAM ECOLOGY AND ASSESSMENT

Purpose & Objectives
Freshwater ecosystems, such as streams, are important sources of biological diversity. It is essential to include physical, chemical, and biological parameters when assessing overall stream health. After completing this exercise the student should be able to:

1. Apply water quality sampling techniques for determining physical, chemical, and biological characteristics of a stream.
2. Investigate the effects of reduced water quality on aquatic life.
3. Explain factors that can lead to reduced water quality.

Introduction
Water is critical to sustaining life on Earth. Therefore, it should not come as a surprise that maintaining stream health is a vital part of sustaining biological diversity in these aquatic ecosystems. Assessing the overall water quality in any aquatic ecosystem is complex as both natural events (i.e., floods, tornadoes, hurricanes) and human inputs (i.e., thermal pollution, sewage, fertilizer runoff) can cause changes to these systems. However, there are some standards scientists use when evaluating water quality and overall stream health. Since the physical and chemical characteristics of the stream will determine the biological diversity, it is critical that all three indicators (physical, chemical, and biological) be measured when assessing water quality. In this investigation we will be focusing on the physical, chemical, and biological indicators of stream health for a flowing water system (lotic) as opposed to a still water system (lentic) such as a pond or lake.

Physical and Chemical Indicators
Evaluating the physical and chemical characteristics of water quality can suggest what type of organisms will be present in that aquatic system. Common physical characteristics included in water quality assessments are velocity of water current, temperature, pH, and total suspended solids in the water column. Typical chemical characteristics included in water quality assessments are: dissolved oxygen, nitrogen compounds, and total soluble phosphates.

Biological Indicators
Many of the organisms living in streams are microscopic algae and protozoa. These organisms are prevalent in streams as they form the base of the food web in this ecosystem. However, due to their size, they can be very difficult to study when assessing water quality. Instead, we will focus on the presence of benthic, macroinvertebrates—organisms lacking a backbone that live on the bottom of the stream bed and can typically be seen without a microscope. This group includes snails, crustaceans, worms, and aquatic insects. Most aquatic insects found will be in the immature, larval form known as a nymph stage (analogous to the caterpillar stage of a moth or butterfly). Macroinvertebrates are highly valuable when assessing water quality, since they are common in most temperate regions of the world and are easy to collect. Aquatic organisms typically live in a given range of tolerance for both physical and chemical characteristics (See Table 10.1). Organisms that live in a narrow range of tolerance are typically the first to disappear if water quality changes and can serve as indicators of water pollution.

Table 10.1 Benthic Macroinvertebrates

Sensitive	Moderately Tolerant	Pollution Tolerant
Stoneflies	Damselflies	Midgeflies
Water Penny Beetles	Dragonflies	Worms
Mayflies	Crayfish	Leeches
Dobsonflies	Amphipods	Pouch Snails
Alderflies	Blackflies	
Mussels	Caddisflies	
Snipeflies	Isopods	
	Craneflies	

From: United State's EPA: Biological Indicators of Watershed Health

Materials

Physical Testing
- pH probe or test kit
- thermometer
- float (for measuring current)
- 50 or 100 meter measuring tape
- 1 liter flask with stopper
- glass fiber filter disk
- graduated cylinder
- small stainless steel or aluminum dish
- desiccators
- water filter apparatus
- balance
- oven

Chemical Testing
- dissolved oxygen probe or test kit
- nitrogen test kit
- phosphate test kit

Biological Testing
- leaf pack bags (can also use empty potato or oranges bag)
- string
- rubber bands
- kick-net – optional
- microscope or stereo dissecting scope
- sterile water
- tryptone glucose extract agar

- warm water bath
- lactose broth
- incubator for bacterial samples
- 9 sterile pipettes
- 4 sterile petri dishes
- 9 fermentation tubes containing lactose broth and small inverted vial
- 2 sterile bottles with stoppers
- sterile flask for water blank

Procedure

Conduct the following water quality tests using water samples taken from two different streams. *Note: If available, it would be best to perform all three of the procedures below in a non-polluted stream and in a potentially polluted stream so students can compare the results.*

1. *Physical Characteristics of the Stream*—perform the following water sampling tests and record results in Table 10.2.

   [Note: All measurements should be taken in mid-stream and at mid-depth. It is best if the section of the stream sampled is free-flowing water and not a stagnant pool on the side of the stream bank.]

   A. Stream Velocity (Current) – Select an area of the stream where it is possible to measure off 20 to 30 meters in length. One student should place a float at the beginning of the distance while the second student standing 20 m to 30m downstream starts a stopwatch. This student will stop the watch once the float reaches him or her. Divide the distance traveled by the time it took to cover the distance (meters/second).

   B. Temperature – Use a thermometer or temperature probe to take a reading of the stream's temperature.

   C. pH – The pH of the stream can be measured using a portable pH meter or by using a water quality test kit.

   D. Total Suspended Solids (TSS) – Use a standard container, such as a 1-liter flask, to capture water from a free-flowing portion of the stream. This can be stoppered and returned to the lab to determine the amount of total suspended solids in the following manner:

      1. Obtain a glass fiber filter disk from the desiccator and weigh it very carefully (to the milligram). Record the weight.
      2. Assemble the filter apparatus as directed by your instructor.
      3. Thoroughly mix the water in your sample to re-suspend any solids that might have settled to the bottom.
      4. Filter 100 ml of the sample
      5. Carefully place the filter on a stainless or aluminum dish and dry in an oven at 103° – 105°C for one hour.
      6. At the end of one hour, cool the filter in a desiccator. Then weigh it and record the weight.
      7. Repeat the drying and weighing cycle until the weight doesn't change.

8. Calculate the amount of suspended solids by using the following formula:

$$\text{Total Suspended Solids (mg/l)} = \frac{(\text{weight of "dirty" filter}) - (\text{weight of clean filter}) \times 1000}{\text{Size of sample filtered (ml)}}$$

2. ***Chemical Characteristics of the Stream***
   Use water quality test kits or sampling probes to perform the following tests. Record your results in Table 10.3.

   A. <u>Dissolved Oxygen</u> – The organisms that live in streams require oxygen, which is dissolved in water, to support aerobic respiration. Some organisms are very sensitive to low oxygen levels, while others are not. Many factors influence the level of dissolved oxygen in an aquatic system.
      1. Factors that increase dissolved oxygen –
         • Photosynthetic organisms such as plants and phytoplankton (algae) release oxygen into the water during daylight hours.
         • Turbulent flow of water increases the ability for oxygen from the air to dissolve into the water.
      2. Factors that decrease dissolved oxygen –
         • Decomposition of organic matter that enters the stream such as runoff that contains human sewage or manure from concentrated animal feeding operations (CAFOs).
         • The death and decay of plants and animals in the water.
         • All organisms (animals, plants, and decay organisms) consume oxygen during respiration. During daylight hours plants carry on more photosynthesis than respiration and release oxygen. At night, plants only carry on respiration and help to deplete oxygen.
         • Thermal pollution from electrical power plants can also cause a drop in dissolved oxygen, since it can increase the temperature of the water, which reduces the solubility of oxygen in water.

   B. <u>Total Soluble Phosphates</u> – Sources that contribute to increased levels of soluble phosphates in water include:
      1. Fertilizer runoff from lawns, golf courses, and agricultural fields.
      2. Organophosphate insecticides and herbicides used to control pests in landscaping and on agricultural fields.
      3. Animal waste from pastures and CAFOs.
      4. Sewage treatment plant effluent.
      5. Household detergents (although many detergent brands in the U.S. have become phosphate-free products).

   C. <u>Nitrogen Compounds</u> – Nitrogen compounds include ammonia (NH_3), nitrate ($-NO_3$), and nitrite ($-NO_2$). Like phosphates, ammonia can make its way to streams from fertilizer runoff, human sewage, and animal waste runoff. Nitrate and nitrite can also be added to aquatic systems by fertilizer runoff or by atmospheric deposition of air pollutants from automobile exhaust.

3. *Biological Indicators in a Stream*

A. Benthic Macroinvertebrates –
1. Collection technique – Take a leaf pack bag and fill it with damp leaf litter from the surrounding stream banks. Fill the bag about half-way full of organic matter and tie it off with a rubber band or string. Submerge the bag in an area of the stream that is flowing and not stagnant. You will need to tie the leaf pack bag to a rock, log, or stake driven into the stream bottom to ensure that it does not float away. Make sure that your leaf pack has as much surface area exposed to the stream as possible and that water can easily flow through it. Leave the leaf pack in the stream for as long as possible—up to two weeks.
If time is limited you may use kick-nets to collect macroinvertebrates. However, leaf pack bags give better results and cause less disruption to the stream.
2. Identification – Empty your leaf pack bag into a shallow tray. Sort through the leaves and rinse them with water as you go. You may find some macroinvertebrates easily and others you may want to find by examining water samples from your tray under a dissecting scope. Identify the organisms by using the identification keys and other resources provide by your instructor. Refer to Table 10.1 to determine if the organisms are sensitive, moderately tolerant, or pollution tolerant. Record your results in Table 10.4.

B. Standard Plate Count – A standard plate count is a method of assessing the total number of several kinds of bacteria in a water sample. Not all bacteria will grow under the conditions used, but most kinds will and a standard plate count gives an index of bacterial numbers.
1. Use a sterile bottle to collect a sample of water from the stream. It is best to collect the water sample in mid-stream and at mid-depth (do not get sediment from the stream bed in the sample). These samples will need to be processed within six hours.
2. To assess the numbers of bacteria present, you will need to prepare a series of dilutions of the original sample as follows (see Figure 10.1):
   a. Obtain a sterile water blank containing 99 ml of water.
   b. Use a sterile pipette to transfer 1 ml of the water sample to the 99 ml sterile water blank. Mix the sample with the water in the water blank.
   c. Obtain four sterile petri dishes labeled 1 ml, 0.1 ml, 0.01ml, and 0.001 ml.
   d. Use a sterile pipette to transfer 1 ml of the *original* water sample to the petri dish labeled 1 ml. Use the following technique to do so. Lift the lid of the petri dish just enough to allow the pipette to deliver the water to the sterile empty dish.
3. Use the same technique to add water to the other petri dishes as follows.
   a. 0.1 ml from the original sample to dish labeled 0.1 ml
   b. 1.0 ml from mixed diluted sample to dish labeled 0.01 ml
   c. 0.1 ml from mixed diluted sample to dish labeled 0.001 ml

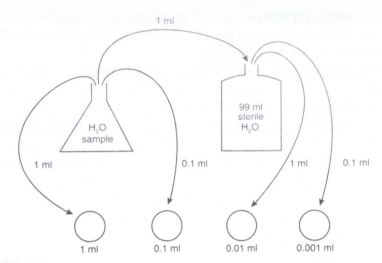

Figure 10.1 Technique for Inocculating Petri Dishes for a Total Plate Count

4. Melt the tryptone glucose extract agar and hold at 44–46°C until used.
5. Lift the lid of each petri dish in turn just high enough to allow you to pour 10–12 ml of the agar into the petri dish. Swirl the contents gently to mix the water sample with the agar.
6. You should also make some petri plates using the sterile water and the medium to see that they were not contaminated.
7. Incubate the petri dishes for 48 hours at 32°C.
8. After 48 hours count the number of bacterial colonies growing on each plate. Since you know the size of the original sample added to each petri dish, you will be able to determine the number of bacteria per millimeter of the original water sample. (Ideally, one of your plates should have been between 30 and 300 colonies. Use this plate for determining the number of bacteria per ml.) Record results in Table 10.5.

C. Coliform bacteria
Coliform bacteria are found in the intestines of humans and other animals; for that reason, the presence of these kinds of bacteria is an indication of contamination from human or animal waste products. The coliform bacteria themselves are not normally a hazard but indicate that other pathogenic (disease-causing) bacteria may also be present. If the source of the coliform bacteria is human, then we can also assume that some human pathogens will be present.

1. Use a sterile bottle to collect a sample of water from both streams—mid-depth, mid-stream. Be careful not to collect any sediment from the bottom when taking the sample. If the sample cannot be used immediately (within one hour), it should be cooled to 10°C or less until it can be used. The sample should be used within six hours of collection.
2. Obtain nine fermentation tubes containing lactose broth. Three should be labeled 10 ml, three should be labeled 1 ml, and three should be labeled 0.1 ml. (See Figure 10.2 below.)
3. Shake the water sample thoroughly.

73

4. Use a sterile pipette to place 10 ml of the water sample into each of the tubes labeled 10 ml; 1 ml into the 1 ml tubes; and 0.1 ml into the 0.1 ml tubes.
5. Incubate at 35°C for 24 hours.
6. After 24 hours, examine the tubes for the presence of gas in the small inverted vial.
7. Examine again at the end of 48 hours.
8. Record all the tubes that show a positive test. A test is positive if gas has collected in the vial *and* the culture is cloudy.
9. Consult a most probable number table to determine the approximate number of coliform bacteria in a 100 ml sample. Your instructor can provide a most probable number table (tables are also provided by the USDA at http://www.fsis.usda.gov/PDF/MLG_Appendix_2_03.pdf). Record results in Table 10.5.

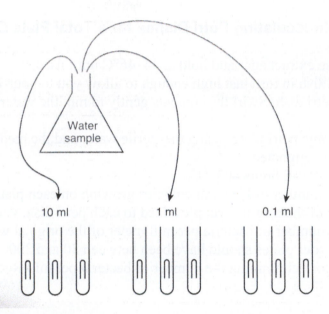

Figure 10.2 Technique for Inocculating Lactose Broth Tubes

EXERCISE 10
STREAM ECOLOGY AND ASSESSMENT

Name:_____

Section:_____

Date:_____

Data Sheet 10.1

Table 10.2 Physical Characteristics

	Stream 1	Stream 2
Stream velocity (meters/second)		
Temperature (°C)		
pH		
Total suspended solids (mg/l)		

Table 10.3 Chemical Characteristics

	Stream 1	Stream 2
Dissolved oxygen (ppm)		
Total soluble phosphate (ppm or mg/l)		
Nitrogen		
Ammonia (ppm or mg/l)		
Nitrate (ppm or mg/l)		
Nitrite (ppm or mg/l)		

Table 10.4 Macroinvertebrate Count

	Stream 1		Stream 2	
	Common Name of Organisms	Number of Organisms	Common Name of Organisms	Number of Organisms
Sensitive Organisms	1. 2. 3. 4.		1. 2. 3. 4.	
Moderately Tolerant Organisms	1. 2. 3. 4.		1. 2. 3. 4.	
Pollution Tolerant Organisms	1. 2. 3. 4.		1. 2. 3. 4.	

EXERCISE 10
STREAM ECOLOGY AND ASSESSMENT

Name:_____

Section:_____

Date:_____

Data Sheet 10.2

Table 10.5 Bacterial Counts

	Stream 1	**Stream 2**
Standard plate count (number/ml)		
Coliform bacteria (number/100ml)		

Analysis

1. Describe three results from your water sampling that differed between the two streams and discuss some possible reasons that could account for these differences.

2. Identify both natural and human factors that would cause
   a. temperature changes in a stream system.

   b. pH changes in a stream system.

   c. increased total suspended solids in a stream system.

EXERCISE 10
STREAM ECOLOGY AND ASSESSMENT

Name:_____

Section:_____

Date:_____

Data Sheet 10.3

3. Compare the results you obtained for the kinds of organisms found in each stream with the data you obtained from your physical and chemical tests. Give specific examples of physical and chemical characteristics of the streams that appear to be correlated with the kinds of organisms found. State how the physical or chemical characteristic affected the organisms.

4. Why might the dissolved oxygen content of a stream be different in the early morning compared with the late afternoon?

5. The U.S. EPA recommends that free-flowing streams have no more than 0.1mg/l of phosphorus to be a healthy stream. Based on this information, is either of your streams healthy?

6. U.S. EPA standard for full body contact (swimming) for coliform bacteria is 200 colonies for 100 ml of water sampled. For less than full body contact (wading) the standard is 2000 colonies for 100 ml sampled. Are your streams safe for swimming or wading? Justify your answer with evidence collected.

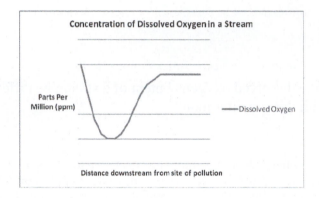

7. Examine the graph above. What are some potential sources of water pollution that would cause a drop in dissolved oxygen? Explain how these inputs cause the overall dissolved oxygen levels to drop.

EXERCISE 11
PERSONAL ENERGY CONSUMPTION

Purpose & Objectives

The purpose of this exercise is to collect specific information about personal energy use, assess the cost of energy use, and identify opportunities to reduce the amount of energy used.

After completing this exercise, the student will be able to:
1. Calculate energy savings from reducing the size of a window, changing the kind of window, or changing the inside temperature of a building.
2. Interpret R-values.
3. Calculate energy loss from a dripping hot water faucet.
4. Examine the implications of an individual's lifestyle on energy consumption.
5. Determine the efficiency of different kinds of light bulbs.
6. Use energy consumption information available to consumers.
7. Determine the energy used by specific electrical devices.
8. Evaluate the long-term cost of purchases of energy consuming products.

Introduction

Except for a few small countries with economies dominated by oil or finance (examples are: Luxembourg, Bahrain, Qatar, and United Arab Emirates), the United States and Canada use more energy per person than all other parts of the world. This is true because historically we have had abundant, inexpensive energy in the form of wood, coal, and oil. Therefore, there was little interest in developing ways to use energy efficiently.

However, as the price of energy has risen there has been considerable interest in improving the efficiency with which we use energy. Government agencies and utility companies provide information about how to improve energy efficiency and also provide financial incentives for doing so. Decisions we make about purchasing and using heating and cooling equipment, water heaters, lighting, transportation, and electrical appliances affect our energy footprint. To a certain extent we can also decide what kind of energy to use (electricity, natural gas, gasoline, diesel, etc.).

There are several complications involved in measuring the quantity of energy contained by various fuels and the amount of energy used by consumers.
1. Fossil fuels and biofuels are mixtures of various chemical compounds and each compound has a specific energy content. For example, gasoline consists of many different derivatives of crude oil but also contains varying amounts of additives, such as ethanol and contaminants, such as water. Therefore, gasoline produced from different sources of crude oil and with different amounts of additives will have different energy content. Similarly, coal and wood contain varying amounts of water and other compounds.
2. The units used to measure different sources of energy differ. Kilowatt-hour, calorie, Btu, joule, and therm are all used as ways to measure the quantity of energy.
3. The United States differs from nearly all other countries because it does not use the metric system as its standard system of measurement. Therefore, a common unit for energy measurement in the United States is the British thermal unit (Btu), but the rest of the world uses the joule (J) as its standard unit for measuring energy. Table 11.1 lists common sources of energy and commonly used values for their energy content. Energy content is given in both Btu, commonly used in the United States, and joules, used in the rest of the world.

Table 11.1 Common Values for the Energy Content of Energy Sources

Energy Source	Commonly Used Units of Measure	Energy Value in Btu	Energy Value in Megajoules (MJ) (1 MJ = 1 million joules)
Gasoline (reg)	Gallon or liter	125,000 Btu/gallon	34.8 MJ/liter
E85 fuel	Gallon or liter	81,800 Btu/gallon	22.6 MJ/liter
Ethanol	Gallon or liter	76,000 Btu/gallon	21.2 MJ/liter
Diesel fuel	Gallon or liter	130,000 Btu/gallon	36.2 MJ/liter
Biodiesel	Gallon or liter	118,300 Btu/gallon	33.0 MJ/liter
Heating oil	Gallon or liter	130,000 Btu/gallon	36.2 MJ/liter
Natural gas	Cubic foot (ft^3) Cubic meter(m^3)	1025 Btu/ft^3	38.2 MJ/m^3
Propane	Gallon or liter	91,500 Btu/gallon	25.5 MJ/liter
Electricity	Kilowatt-hour (kWh)	3414 Btu/kWh	3.6 MJ/kWh
Coal (anthracite)	Pound (lb) Kilogram (kg)	12,700 Btu/lb	29.5 MJ/kg
Wood	Pound (lb) Kilogram (kg)	7870 Btu/lb *Highly variable*	18.8 MJ/kg *Highly variable*

Procedure

This exercise consists of a series of separate activities that examine energy use. Complete each part and record your results on the appropriate Data Sheets.

Note: Your instructor may choose to divide the class into groups and assign specific parts of this exercise to each group. The results can be compiled and shared with all members of the class.

Materials Needed for Each Group

Measuring tape

Two thermometers

Balance

Beaker

Lighting fixture for screw-in light bulbs

Natural gas, electricity, and fuel oil bills

Calculators

Light meters

Exercise 11.1 Heat Loss/Gain through Windows

Heat leaves or enters buildings by passing through walls, ceilings, doors, and windows. Most walls and ceilings contain insulating materials that resist the flow of heat. Windows present a particular problem because glass has the desirable quality of letting in light but an undesirable quality of having a low insulating value (R-value).

The R-value is a measure of how well a material resists the flow of heat through it. The reciprocal of R (1/R) is equal to the number of Btu that would pass through a 1-square-foot (1 ft^2) surface in 1 hour if the difference in temperature on opposite sides of the surface is $1°F$. Therefore, we can calculate heat loss or gain through a window by using the following formula.

$$\text{heat loss/gain in Btu per hour} = \frac{\text{ft}^2 \times \text{difference in temperature (}^\circ\text{F)}}{\text{R-value of the window}}$$

Formula 1

Although glass is a poor insulator, windows can be designed with additional panes with a space between panes and special coatings on the glass that greatly improve the R-value. See Table 11.2.

Table 11.2 Typical R-values for Different Window Designs

Type of Window	United States R-value	International (Metric) R-value
Single-pane glass	0.9	0.16
Double-pane glass (1/2 inch air space)	2.04	0.36
Double-pane low-emission glass (1/2 inch air space)	3.13	0.55
Triple-pane glass (1/2 inch air space)	3.23	0.57
Triple-pane High efficiency	5	0.88

Record all results from steps 1–7 on Table 11.4 Heat Loss/Gain through a Window on Data Sheet 11.1.

1. Choose a window in your classroom or home and measure the height and width of the glass surface in inches. Calculate the surface area in ft^2 by using the following relationship:

    $$\text{Surface area}_____ \text{ft}^2 = \frac{\text{Height}\_\_\_\text{inches} \times \text{width}\_\_\_\text{inches}}{144 \text{ in}^2/\text{ft}^2}$$

2. Measure the temperature on the inside and outside of the window in °F and calculate the difference in temperature between the inside and the outside of the window.
3. Determine the R-value of the window from Table 11.2.
4. Calculate the rate of heat transfer through the window by using **formula 1.** Record this information in the *Heat loss or gain per hour* column of Table 11.4.
5. Calculate the effect on the rate of heat transfer if the window you measured were replaced with a window with a better R-value. Consult Table 11.2 for a list of typical R-values for windows. Use **formula 1** to calculate the heat transfer.
6. Use **formula 1** to calculate the rate of heat transfer if the size of the window were reduced by 50 percent.
7. Use **formula 1** to calculate the effect a 5°F decrease in the temperature on the inside of the window will have on the rate of heat transfer.
8. Calculate an estimate of the yearly heat flow through the window by doing the following:
    a. On the internet, type in "average monthly temperature" and your city and state.
    b. You will get a number of websites that publish the data you want. Choose data for average or mean temperature for each month. Enter this information into the *Mean Monthly Temperature* column of Table 11.3 below.
    c. Assume that an average temperature above 65°F requires cooling (air conditioning) and that an average temperature below 65°F requires heating and complete Table 11.3.

Table 11.3 Calculating Annual Energy Demand for Heating and Cooling

Month	Mean Monthly Temperature (°F)	Difference between Mean Monthly Temperature and 65°F*		Number of Hours in a Month	Total Degrees Above or Below 65°F
Example	*30*	*65–30=*	*35*	*× 720 =*	*25,200*
January		65–___=		× 744 =	
February		65–___=		× 672 =	
March		65–___=		× 744 =	
April		65–___=		× 720 =	
May		65–___=		× 744 =	
June		65–___=		× 720 =	
July		65–___=		× 744 =	
August		65–___=		× 744 =	
September		65–___=		× 720 =	
October		65–___=		× 744 =	
November		65–___=		× 720 =	
December		65–___=		× 744 =	
Total number of degrees above or below 65°F per hour					

*In months that have a mean temperature above 65°F your result will be negative (65–70 = –5). Ignore the negative sign and treat the number as if it were positive because energy will be flowing into the building and air conditioning, which requires energy, will be needed to get rid of the heat.

d. Use the following formula to calculate an estimate of the yearly flow of heat through the window you measured. Use the same number of ft² and the R-value for the window but insert the total number of degrees above or below 65°F for difference in temperature. Insert the result of your calculation into the *Heat loss or gain per year* column of Table 11.4. Heat Loss/Gain through a Window on Data Sheet 11.1.

$$\text{heat loss/gain in Btu per year} = \frac{\text{ft}^2 \times \text{difference in temperature (°F)}}{\text{R-value of the window}}$$

e. Perform the same calculation for the *High R-value window* and *Reduce window surface area by 50%* rows of Table 11.4. Heat Loss/Gain through a Window.

f. Determine the annual cost of the energy passing through the window in the following manner:

    i. Determine the unit cost for the energy sources used to heat or cool your home. Depending on the energy sources used, the unit cost will be expressed as cents/kWh, cents/ft^3, or $/gallon.

    NOTE: You will probably need to determine the unit cost of more than one energy source, since you may use natural gas, propane, or fuel oil for heating but will probably use electricity for air conditioning.

    CAUTION: Utility bills for electricity and natural gas use are often misleading. In addition to the actual cost of electricity or natural gas consumed, there are distribution charges, fees, surcharges, taxes, and other charges. Therefore, in order to obtain the actual cost of a kilowatt-hour (kWh) of electricity it is necessary to divide the total of all costs associated with electrical use by the number of kilowatt-hours used. Similarly, to obtain the actual cost for a unit of natural gas (cubic foot or cubic meter) you need to divide the total cost associated with the use of natural gas by the number of units of natural gas used.

$$\frac{\text{Unit cost}}{\text{(cents/kWh or cents/ft}^3 \text{ or \$/gallon)}} = \frac{\text{Total cost in dollars}}{\text{Units used}}$$
$$\text{(kWh or ft}^3 \text{ or gallons)}$$

    ii. Determine the number of Btu of energy contained in a unit of fuel. Table 11.1 has the typical Btu content of commonly used fuels.

    iii. Divide the total heat gain or loss (Btu) by the number of Btu in a unit of fuel (Btu/kWh or Btu/ft^3 or Btu/gallon).

    iv. Multiply the number of units of fuel used to provide the heating or cooling (kWh or ft^3 or gallons) by the unit price for the fuel (cents/kWh, cents/ft^3, $/gallon, etc.)

    v. Record your results in the *Cost of heat loss or gain per year* column of Table 11.4 on Data Sheet 11.1.

Exercise 11.2 Heating Water

Compared with other common substances, water has a high specific heat, which is the amount of heat energy needed to change the temperature of a given mass of a substance by 1°C. This means that water resists changes in temperature. In other words, it takes a lot of heat energy to make a small change in the temperature of water. Therefore, water heaters are quite expensive to operate. It takes 1 Btu of heat energy to raise the temperature of 1 pound of water 1°F. In this exercise we will look at the costs associated with a leaking hot water faucet.

1. Assume you have a slow leak in a hot water faucet.
2. Turn on a water faucet so that it is leaking at a rate of about one drip per second.
3. Capture this water for a period of fifteen minutes.
4. Weigh the amount of water you have collected (in pounds). Enter this result in Table 11.5 Heating Water on Data Sheet 11.1. Use Table 11.5 as you do the following calculations.

    i. Multiply the number of pounds of water collected by 35,040 (There are 35,040 fifteen minute periods in a year.) to find the amount of water that would be lost in one year.

    ii. Assume that water enters the water heater at 40°F and leaves the water heater at 120°F, a temperature difference of 80°F. Multiply the pounds of water lost in one year by 80 to obtain the number of Btu of heat energy that would be lost in one year if the leak were not fixed.

5. Calculate the number of kilowatt-hours of electricity or cubic feet of natural gas it would take to produce the heat wasted by not fixing the dripping faucet. (Use Table 11.6 Cost to Heat Water on Data Sheet 11.1 as you do steps 5 and 6.)

$$\text{Kilowatt-hours} = \frac{\text{———Btu}}{3414 \text{ Btu / kWh}}$$

$$\text{ft}^3 \text{ of natural gas} = \frac{\text{———Btu}}{1025 \text{ Btu/ft}^3}$$

6. Calculate the cost associated with this water loss by multiplying the number of kilowatt-hours of electricity or cubic feet of natural gas by the cost per kWh or ft^3 determined from your utility bill.

Exercise 11.3 Lighting

Lighting is something that we take for granted. We usually simply flip a switch and it is instantly there. However, what does it cost in energy to provide the light we use and does the type of light source make a difference in that energy cost?

1. In a dark room, hold a light meter exactly 10 feet from a 40-watt incandescent light bulb.
2. Record the light meter reading on Table 11.7 on Data Sheet 11.2.
3. Use the light meter to measure the amount of light coming from a 40-watt fluorescent light bulb.
4. Since both bulbs have the same wattage, they use the same amount of electrical energy in the production of light. The light-meter reading will be in foot-candles, LUX, or lumens. It doesn't really matter which measuring system is used, since you are only going to compare the two light-meter readings with each other.
5. Calculate the number of units of light produced per watt of energy consumed. Record your results on Table 11.7 Lighting on Data Sheet 11.2. Which of the bulbs produces the most light per watt?
6. You may want to compare other kinds of light bulbs (halogen, LED).
7. Record your results on Table 11.7 Lighting on Data Sheet 11.2.

Exercise 11.4 Transportation

North Americans look at freedom of movement as a right. We drive and fly more than any other people in the world. In some urban areas, public transport such as trains, subways, and buses provide efficient ways to travel about the city, reducing traffic jams and air pollution.

1. For one week, keep a log of all the miles you travel by the following methods:
    a. Foot
    b. Bicycle
    c. Automobile
    d. Public transport (train, subway, bus, trolley, etc.)
    e. Plane
    f. Other
2. Approximately what percentage of each of these trips was required for work, school, or other necessary reasons?
    a. Foot
    b. Bicycle
    c. Automobile
    d. Public transport (train, subway, bus, trolley, etc.)
    e. Plane
    f. Other
3. Record your data on Table 11.8 Transportation, on Data Sheet 11.2.

Exercise 11.5 Electrical Appliances

Electrical appliances are very convenient. They allow us to do things quickly and relieve us of distasteful or tedious tasks. How much energy do you use as a result of such devices? It is important to recognize that the total energy cost of an appliance also includes the energy necessary to manufacture, distribute, and merchandise the item. However, it might be instructive to determine how much energy is consumed by the use of the various electrical appliances. You can find the wattage of an electrical appliance on a label on the appliance.

1. Keep a log of all the electrical appliances you use in a one-week period. List each appliance and the number of minutes it was used per week. Record your data and do your calculations on Table 11.9 Electrical Appliances on Data Sheet 11.3.
2. Determine the wattage of the appliance from its label. If the label does not show the wattage but gives volts and amperes, you can calculate wattage as follows:

$$Watts = Volts \times Amperes$$

Record your results on Table 11.9 on Data Sheet 11.3.
3. If you know the wattage and the number of minutes it was used, you can calculate the number of kilowatt-hours of energy used (see the equation below).

$$Kilowatt\text{-}hours\ used = \frac{(watt\ rating) \times (total\ minutes\ used)/60}{1000}$$

Record your results on Table 11.9 on Data Sheet 11.3.
4. Calculate the cost of the energy used. (Cost = kWh used × cents/kWh)
Record your results on Table 11.9 on Data Sheet 11.3.

Exercise 11.6 Long-term Costs of Energy Using Appliances

When you make the decision to purchase an appliance one of the primary things you look at is the purchase price. However, in addition to the purchase price, the long-term cost of that decision includes the amount of energy the appliance will use. The most energy intensive appliances are those that are involved in heating and cooling—furnaces, air conditioners, refrigerators, freezers, and water heaters.

Choose one of these appliances and visit a local appliance store to obtain information about the cost and energy usage of the appliance. Look at the EnergyGuide label on the appliances and choose two appliances with similar features but with significantly different energy consumption. See Figure 11.1 for an example of an EnergyGuide label.

Show your data and calculations on Table 11.10 Long-Term Cost of Appliances on Data Sheet 11.3.

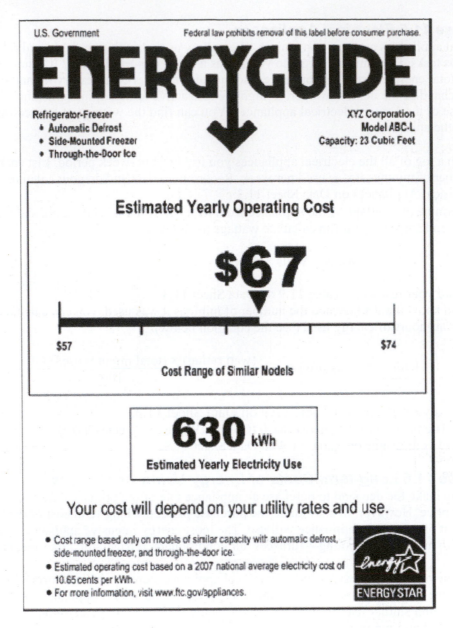

Figure 11.1 An EnergyGuide Label

EXERCISE 11
PERSONAL ENERGY CONSUMPTION

Name:_____

Section:_____

Date:_____

Data Sheet 11.1

Table 11.4 Heat Loss/Gain through a Window

Window Characteristics	Surface Area (ft²)	Inside Temp. (°F)	Outside Temp. (°F)	Temp. Difference (°F)	R-Value	Heat Loss or Gain per Hour (Btu)	Heat Loss or Gain per Year (Btu)	Cost of Heat Loss or Gain per Year ($)
Low R-value window								
High R-value window								
Reduce window surface area 50%								
Reduce inside temperature 5°F								

Table 11.5 Heating Water

	Pounds of Water Collected in 15 min		Pounds of Water Lost per Year		Btu Needed to Heat the Water Lost in a Year
Dripping Faucet	39.9 lb	X 35,040 =	1,398,096	X 80 =	111,847.68

Table 11.6 Cost to Heat Water

	Btu Used	kWh/Btu or ft³/Btu		Price per kWh or ft³	Cost
Electricity	1064	÷ 3414 Btu/kWh =	.3117 kWh	× .22 cents/kWh =	.68
Natural gas		÷ 1025 Btu/ft³ =	_____ ft³	× _____ cents/ft³ =	

EXERCISE 11
PERSONAL ENERGY CONSUMPTION

Name:_____
Section:_____
Date:_____

Data Sheet 11.2

Table 11.7 Lighting

Kind of Light Bulb	Light Meter Reading (lumens)	Light/Watt (Divide meter reading by number of watts.)
40 watt incandescent		
40 watt fluorescent		
40 watt halogen		
Other		

Table 11.8 Transportation Data

Miles per day									
Mode of transport	Sun	Mon	Tue	Wed	Thurs	Fri	Sat	Total	% Necessary
Walking	1.2 mi	2.7 mi	3.1 mi	3.1 mi	2.4 mi	2.6	1.7	15.1 mi	8.15%
Bicycle									
Car	7 mi	4	5	18	4	7	13	58 mi	
Public transport									
Plane									
Other									

EXERCISE 11
PERSONAL ENERGY CONSUMPTION

Name:_____
Section:_____
Date:_____

Data Sheet 11.3

Table 11.9 Electrical Appliances

Appliance	Wattage (volts × amps)	Minutes Used per Week	kWh used (Watts/1000 × min/60)	Cost (kWh × price/kWh)
Television	116	630	1.218	0.26796
Desktop or laptop computer	70	1050	1.225	0.2695
Sound system	150	210	0.525	0.1155
Air conditioner	630	2100	22.05	4.851
Clothes drier				
Hair drier				
Shaver	15	30	0.0075	0.00165
Microwave	600	200	2	0.44
Washer				

Table 11.10 Long-term Costs of Appliances

	Purchase Price	Energy Used per Year (kWh or ft³)	Annual Cost of Energy (cents/kWh or cents/ft³)	Ten Year Cost of Appliance (annual cost of energy × 10 + purchase price)
Low energy consuming appliance				
High energy consuming appliance				

EXERCISE 11
PERSONAL ENERGY CONSUMPTION

Name:_____

Section:_____

Date:_____

Data Sheet 11.4

Analysis

1. How much energy would you save by replacing the low efficiency window with a high efficiency window? What are the $ savings per year?

2. What effect does reducing the size of a window have on the amount of heat gain or loss?

    Why does this change occur?

3. What effect does lowering the inside temperature by 5°F have on heat gain or loss?

    Why does this change occur?

4. Why does lowering the temperature in a house at night during the winter save energy?

5. Why does raising the temperature in a house during the summer save energy?

6. Why does lowering the temperature setting of your water heater save energy?

EXERCISE 11
PERSONAL ENERGY CONSUMPTION

Name:_____

Section:_____

Date:_____

Data Sheet 11.5

7. The U.S Congress passed a law to make incandescent light bulbs obsolete. Is this a good idea from an energy point of view? Why or why not?

8. What percent of your weekly travel could have been avoided?

9. At current energy prices, which are least expensive to operate, gas or electric appliances? Is the difference significant?

10. If you had an old refrigerator that used twice as much electricity as the best model you researched and you replaced it with the more efficient model, how many years would it take for you to recover the cost of the new refrigerator? Calculate this by dividing the purchase price of the new refrigerator by the number of dollars saved per year because of the lower energy costs.

11. Many electric utility companies pay rebates to customers who replace old refrigerators, freezers, heating systems, or poorly designed windows. How does the utility company benefit?

12. What percentage of the purchase price of an automobile is the cost of fuel for one year if you were to drive 20,000 kilometers (12,000 miles) per year?

EXERCISE 12
INSULATING PROPERTIES OF BUILDING MATERIALS

Purpose & Objectives

This exercise is designed to help students gain an understanding of how the insulating value (R-value) of a material is determined and to compare the insulation value of different building materials. After completing this exercise the student will be able to:

1. Test the insulating effectiveness of various kinds of building materials.
2. State how R-values are determined.
3. Describe the importance of understanding the R-values of building materials to designing energy efficient buildings.

Introduction

As energy resources become scarce, prices rise and ways are sought to reduce energy use. One of the most economical ways to reduce heat loss or gain in buildings is by using good insulation and energy efficient windows in construction. It is also possible to install insulation in already constructed buildings and replace inefficient windows to reduce heat loss or gain. In the United States, the standard unit used to describe insulating value is an R-value, a material's ability to resist the flow of heat through it. Materials with a high R-value have a high insulating ability. The R-value has the units R = ft²•°F•h/Btu. 1/R (the reciprocal of R) is a measure of the amount of heat energy in British thermal units (Btu) that would pass through a piece of material 1 square foot in area in 1 hour when the temperature is 1° Fahrenheit higher on one side of the insulation than on the other. Therefore, you can use the following formula to calculate heat loss or gain.

$$\text{heat loss/gain in Btu per hour} = \frac{\text{ft}^2 \times \text{difference in temperature (°F)}}{\text{R-value}}$$

A Btu is the amount of heat energy necessary to raise 1 pound of water 1°F (from 39°F to 40°F). It is a very small amount of energy equal to about the amount of heat produced by burning about ½ a drop of gasoline or about ½ a cubic inch of natural gas.

In countries that use the metric system the R-value is determined in the same basic manner but the area is measured in square meters (m²), the temperature is in degrees Celsius (°C), and rate of energy flow is in watts (W). Therefore, the R value has the units R = m²•°C/W. Unfortunately in both systems the rate of heat gain or loss is called R but the numerical value is different. The U.S. R-value is 5.678 times larger than the metric equivalent. To convert a U.S. R-value to a metric R-value you must divide the U.S. R-value by 5.678. (A U.S. R-value of R = 30 would have a metric equivalent of R = 5.3.) Table 12.1 lists typical United States and metric R-values for several kinds of materials.

Table 12.1 R-Values of Common Materials Used in Housing

Material	U.S. R-value	Metric R-value
No insulation	0	0
4-inch brick	0.44	0.08
Asphalt shingle	0.44	0.08
½-inch gypsum board (dry wall)	0.45	0.08
Vinyl or aluminum siding (uninsulated)	0.61	0.11
½-inch plywood	0.63	0.11
Single-pane glass	0.9	0.16
Double-pane glass (½-inch air space)	2.04	0.36
Double-pane low-emission glass (½-inch air space)	3.13	0.55
1-inch wood	1.4	0.25
1-inch fiberglass batting	3.1–4.3	0.55–0.76
1-inch extruded polystyrene board	5–5.4	0.88–0.95
1-inch closed cell polyurethane foam board	7–8	1.23–1.41
1-inch foil-faced polyisocyanurate	7.2	1.27

Procedure

1. You will use a simple apparatus to test and compare the relative insulating ability of several materials used in building construction. The apparatus consists of two wooden boxes with inside dimensions of 8.5 inches high by 8.5 inches wide by 18 inches long. One end of each box is left open. One box contains a light bulb that, when turned on, will heat the air. The second box will have a thermometer inserted into the top of it.

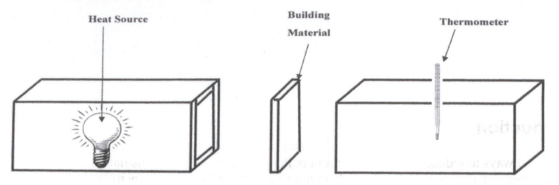

Figure 12.1 Apparatus for measuring insulation efficiencies of different materials

2. Your instructor will divide the class into groups and assign the materials whose insulating properties will be tested. In order to have a basis for comparison (control), at least one trial should be run with no insulating material in the apparatus.

3. Your instructor will provide examples of various materials used in construction of buildings. These materials will be placed between the two open ends of the boxes and the boxes will be pushed tightly against the building material. Depending on the insulating properties of the material placed between the two boxes, different amounts of heat will be transferred from the box with the light bulb through the building material to the second box (see Figure 12.1).

    Possible building materials to test:
    >   Single-pane glass
    >   Double-pane glass
    >   Triple-pane glass
    >   Low-E glass
    >   ½-inch thick plywood
    >   1-inch thick fiberglass batting
    >   1-inch thick extruded polystyrene
    >   Aluminum foil
    >   ½-inch thick gypsum board (dry wall)
    >   Asphalt shingle

4. Turn on the light. Use the thermometer to note the temperature at the beginning of the exercise and at 5-minutes intervals for 30 minutes. Record your measurements on Table 12.2 on Data Sheet 12.1. You will be able to compare the insulating properties of the various materials tested by comparing the rate of temperature change. The greater the insulating value of the material, the less the temperature will rise in the second box.

5. Share your data with the class.

6. Plot the data on Graph 12.1 on Data Sheet 12.1 or alternatively insert the data into an Excel spreadsheet and use the scatter (x-y) graph option to construct a graph of the class data. In Excel you can also apply a linear trend line to your data.

EXERCISE 12
INSULATING PROPERTIES OF
BUILDING MATERIALS

Name:_____

Section:_____

Date:_____

Data Sheet 12.1

Analysis

Table 12.2 Temperature Change with Different Construction Materials

Time (Minutes) / Insulating Material	Record the temperature at the beginning (Time 0) and at 5-minute intervals						
	Control	Material 1	Material 2	Material 3	Material 4	Material 5	Material 6
0							
5							
10							
15							
20							
25							
30							

Graph 12.1 Change in Temperature

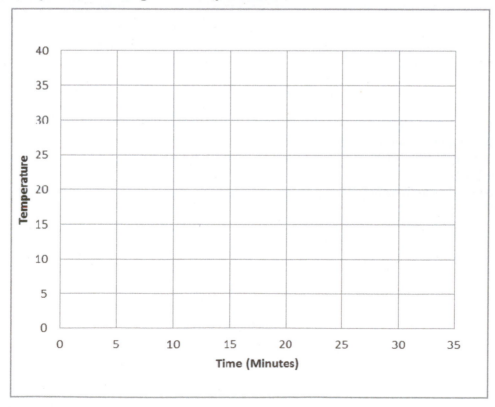

95

| EXERCISE 12 INSULATING PROPERTIES OF BUILDING MATERIALS | Name:_____ Section:_____ Date:_____ |

Data Sheet 12.2

Analysis

1. Which of the materials tested provided the most effective insulation? How do your data support this conclusion?

2. If you had to choose the thinnest insulation to use, based on your results, which insulation would you use and why?

3. Although you did not directly measure R-values, compare your results for the flow of heat through the building material with published R-values for the materials you tested. Are your results consistent with the R-values? Use the data you collected to support your conclusion.

4. A wall typically consists of several layers of building materials. Beginning with the inside there is usually:
    a) Dry wall (0.5–0.75 inch thick)
    b) Space filled with insulation (3.5–5.5 inches thick)
    c) Wood product sheeting such as plywood, particle board, etc. (0.75 inches thick)
    d) Siding or brick
    Each of these products has an R-value. What percent of the insulation value of a wall is the result of the insulation?

5. Many new homes are built with walls that are thicker than conventional homes. The space that the insulation occupies is 2 inches thicker. How much would this increase the R-value of the wall?

EXERCISE 13
THE EFFECTS OF RADIATION ON THE GERMINATION AND GROWTH OF SQUASH SEEDS

Purpose & Objectives

This exercise is designed to show the effects of different amounts of radiation on the germination and growth of squash seeds. After completing this exercise the student will be able to:

1. Determine the effect of radiation on the time of germination of squash seeds.
2. Calculate the effect of different amounts of radiation on germination percentages.
3. Quantify the effect of different amounts of radiation on the growth of the roots of irradiated squash seeds.
4. Estimate threshold levels of radiation on time of germination, ultimate germination percentage, and root growth.

Introduction

Gamma radiation is a form of energy similar to X-rays, which along with other forms of radiation, is emitted from a variety of environmental sources. Two other kinds of radiation are beta radiation and alpha radiation. Beta radiation consists of rapidly moving electrons, and alpha radiation consists of rapidly moving particles that are composed of two protons and two neutrons. Atoms of the same element that differ from one another in the number of neutrons present are called *isotopes*. Some isotopes of atoms, such as cobalt-60, are unstable and release radiation when the nucleus of the atom disintegrates. Radiation is of concern because, depending on its type, total amount, or rate of delivery, radiation can cause changes in the genetic material (DNA) within cells or change the activities of cells. In very high doses, it kills cells directly.

In the United States there are two units commonly used to quantify the amount of radiation encountered by objects, the *rad* and the *rem*. A **rad** (*r*adiation *a*bsorbed *d*ose) is a measure of the amount of radiation absorbed (absorbed dose) and is equal to 0.01 joules of energy absorbed per kilogram of material (0.01 joule/kilogram). Most human radiation doses are measured in millirads (mrad). One rad = 1000 mrads.

The **rem** (*r*oentgen *e*quivalent *m*an) is a measure of the amount of biological damage that can be done by a given absorbed dose. The rem is called a dose equivalent because it is used to measure the amount of damage that can be done by a rad of different kinds of radiation. A rem is equal to a rad times a quality factor. Each of the kinds of radiation has a specific quality factor. Alpha radiation has a quality factor of 20. Beta radiation and gamma radiation have a quality factor of 1.

Unfortunately, the United States is out of step with the rest of the world in terms of the units used to measure radiation exposure. The rest of the world uses the gray (Gy) to quantify the absorbed dose and the Sievert (Sv) to quantify the dose equivalent. The Gy and Sv are 100 times larger than the rad and rem. (100 rads = 1 Gy and 100 rems = 1 Sv.)

In this exercise you will evaluate the effects of gamma radiation on squash seeds. The seeds have received the following amounts of gamma radiation from a cobalt-60 source:

0.1 rads (Unirradiated Control—background radiation)
50,000 rads
150,000 rads
500,000 rads
4,000,000 rads

Table 13.1 gives common radiation exposures and the known effects in humans.

Table 13.1 The Effects of Some Common Radiation Exposures*

Kind of Exposure	Dose (U.S.)	Dose (International)	Effects of Exposure (at dose shown)
Television Set	0.5 mrem/hour	0.005 mSv/hour	None known
High Altitude Plane Flight	0.5 mrem/hour	0.005 mSv/hour	None known
Chest X-ray	10 mrem/exposure	0.1 mSv/hour	None known
Mammogram	30 mrem/exposure	0.3 mSv/hour	None known
Natural Background	100–200 mrem/year	1–2 mSv/year	None known
Allowable Maximum Exposure for Public	500 mrem/year	5 mSv/year	None known
CT Scan	1,100 mrem/exposure	11 mSv/exposure	None known
Allowable Exposure for Nuclear Facility Worker	5,000 mrem/year	50 mSv/year	None known
Exposure of Embryo	10,000 mrem/exposure	100 mSv/exposure	Embryo abnormalities possible
Nuclear Accident	75,000 mrem/incident	750 mSv/incident	Radiation sickness in some people
Nuclear Accident	400,000 mrem/incident	4,000 mSv/incident	50% of people exposed will die

*Please note that for gamma radiation the quality factor is 1, so the absorbed dose (rad) is the same as the dose equivalent (rem). Also note that the radiation dose to the squash seeds is huge. The smallest dose of 50,000 rads is equal to 50,000,000 mrads (or mrems), which would kill any exposed human.

Procedure

1. Obtain 5 petri dishes with lids.

2. Cut paper towels to fit the bottom (smaller of the two halves) of the petri dish.

3. Use a wax pencil or piece of tape and label the bottom of the petri dishes with your name, date, and the radiation dose the seeds received as follows:
   a. Control
   b. 50,000 rads
   c. 150,000 rads
   d. 500,000 rads
   e. 4,000,000 rads

4. Wet the paper towels in the bottom of the petri dish. The towels should be very wet not just damp.

5. Into each of the 5 labeled petri dishes, place ten (10) seeds of the appropriate kind. They should be spread out on the paper towel.

6. Examine the seeds every 24 hours or at times specified by your instructor and record the number of seeds that have germinated in each of the dishes. The seeds will begin to germinate in 2-4 days depending on the temperature. Seeds that have just germinated will show a tiny, white growth (the root) protruding from the pointed end of the squash seed. Record the data on Table 13.2 Effect of Radiation Dose on Time of Germination on Data Sheet 13.1.

7. After 7 days or a period of time specified by your instructor,
    a. Record the total number of seeds that germinated in each petri dish.
    b. Measure in millimeters the length of the root of each of the seeds that germinated and calculate the average length of roots for each of the radiation doses. You will need to remove the plants from the dishes and use a millimeter ruler to do this. Since some shoots will be curved, you will need to gently straighten them to get an accurate measurement of length.
    c. Record these data on Table 13.3 Effect of Radiation Dose on Germination and Root Length on Data Sheet 13.1.

8. At this point your instructor may have you combine the data collected by all students to create combined versions of Tables 13.2 and 13.3.

9. On Data Sheet 13.2 use the data from Table 13.2 to construct Graph 13.1, showing the number of seeds that germinated versus time. You will have five different lines (one for each of the radiation doses) on this graph, so make sure that you label each.*

10. Use the data from Table 13.3 to construct Graph 13.2 (Maximum Number of Seeds Germinated versus Radiation Dose) and Graph 13.3 (Average Root Length versus Radiation Dose) on Data Sheets 13.2 and 13.3. Because there is a wide range of dosages, it is convenient to plot the log of the dosage.*

    *You may want to enter the data from Tables 13.2 and 13.3 into separate Excel databases and use the graphing capabilities of Excel to generate the graphs. You will need to use the scatter (x-y) plot. For Graphs 13.2 and 13.3 you will also need to change the dose axis to a log scale.

EXERCISE 13
THE EFFECTS OF RADIATION ON THE
GERMINATION AND GROWTH OF SQUASH SEEDS

Name:_____

Section:_____

Date:_____

Data Sheet 13.1

Table 13.2 Effect of Radiation Dose on Time of Germination

Treatment / Day	Number of Seeds Germinated				
	Control (0.1 rads)	50,000 rads	150,000 rads	500,000 rads	4,000,000 rads
0	0	0	0	0	0
1					
2					
3					
4					
5					
6					
7					

Table 13.3 Effect of Radiation Dose on Germination and Root Length

Dose	Control (0.1rads)	50,000 rads	150,000 rads	500,000 rads	4,000,000 rads
Maximum Number Germinated					
Average Root Length (mm)					

EXERCISE 13
THE EFFECTS OF RADIATION ON THE
GERMINATION AND GROWTH OF SQUASH SEEDS

Name:_____
Section:_____
Date:_____

Data Sheet 13.2

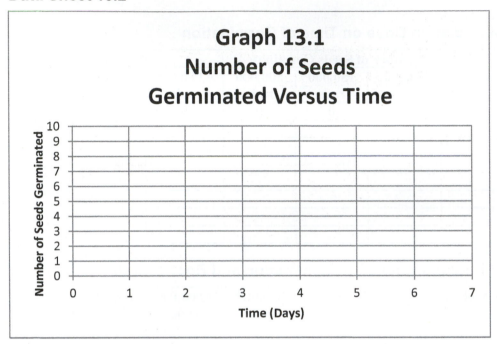

Graph 13.1
Number of Seeds
Germinated Versus Time

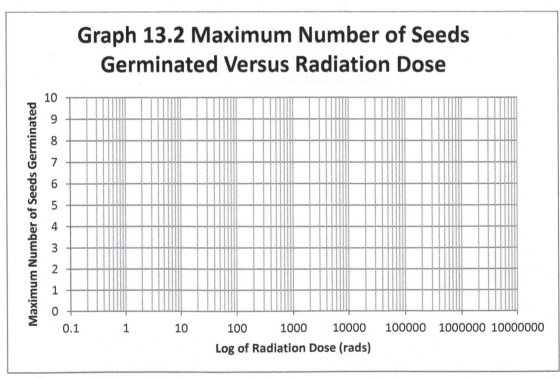

Graph 13.2 Maximum Number of Seeds
Germinated Versus Radiation Dose

EXERCISE 13
THE EFFECTS OF RADIATION ON THE
GERMINATION AND GROWTH OF SQUASH SEEDS

Name:_____
Section:_____
Date:_____

Data Sheet 13.3

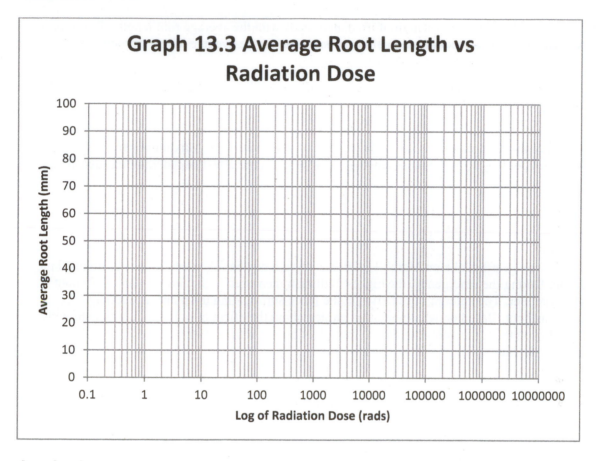

Analysis

1. Compare the time of germination of irradiated seeds to the control.
    At what dose do you recognize an effect of radiation on the time of germination?

2. Compare the final number of seeds germinated for each dose with the control.
    At what dose do you recognize an effect of radiation on the ultimate number of seeds that germinated?

EXERCISE 13
THE EFFECTS OF RADIATION ON THE
GERMINATION AND GROWTH OF SQUASH SEEDS

Name:_____

Section:_____

Date:_____

Data Sheet 13.4

3. Compare the average root length of irradiated seeds with the average root length of the control. At what dose do you recognize an effect of radiation on the average length of roots?

4. Which of the three comparisons made (1.Time of germination, 2. Maximum germination, or 3. Average root length) is the most sensitive way to detect the effect of radiation on squash seeds?

5. The massive earthquake on March 11, 2011, caused a nuclear plant emergency at Fukushima, Japan. During the emergency, the highest radiation measured in the power plant was about 40 rems/hour. The amount of radiation at the fence surrounding the plant was about 0.2 rems/hour. Based on what you have learned about the effect of radiation on squash seeds, how do you think the germination and growth of plants in the vicinity of the Fukushima nuclear accident in Japan will be affected?

6. Are the effects of radiation on plants a good way to predict the effects of radiation doses on humans? Defend your position.

7. A common problem associated with monitoring the effects of radiation on living things is to determine the level below which there is no detectable effect. This is known as the threshold level. Based on the results you obtained during this exercise, what is your estimate of the threshold level below which there is no detectable effect on the germination or growth of squash seeds? How does this compare to the published data on human exposures shown in Table 13.1?

EXERCISE 14
EVALUATING RENEWABLE ENERGY SOURCES

Purpose & Objectives

There are conflicting claims about the cost effectiveness and environmental effects of utilizing various renewable energy sources. In this exercise you will use information from the National Renewable Energy Laboratory (NREL) and other sources to gain insight into some of the issues related to renewable energy development. After completing this exercise the student will be able to:

1. Access information from the NREL website.
2. Determine the potential for wind energy development in their local area and their state.
3. Calculate the solar energy potential for their locality.
4. Evaluate the cost of ethanol fuel.
5. Evaluate the potential for using currently available biomass for energy production.
6. Evaluate the cost of producing electricity from various renewable energy sources.

Introduction

There has been increased interest in renewable energy sources in recent years as the cost of fossil fuels has risen and continued political strife in the Middle East makes world sources of oil uncertain. The primary sources of renewable energy are biomass, solar energy, wind, geothermal energy, and hydroelectric power. Biomass and hydroelectric power are the most developed of the renewable technologies. Great amounts of money and much manpower have been expended by both government and private companies to develop these technologies. Renewable energy sources provided nearly 13% of electricity in the United States in 2011.

However, there are issues related to the further development of renewable energy. In countries where solar and wind energy have been expanded greatly (United States, Germany, Japan) there have been substantial government subsidies provided to encourage the purchase and deployment of these technologies. In many cases, the further development of wind and hydroelectric resources have met with resistance from the public.

Although currently biomass is the largest source of renewable energy, further development raises some issues. Biomass that is removed from cropland and forests does not contribute to the structure and fertility of the soil. In addition, sources of biomass are often located far from centers of population, which means that transportation of these materials is costly.

While there is great potential for the development of hydroelectric power in the world, most of the suitable hydroelectric sites in the United States have already been developed. Therefore, this exercise will not deal with hydroelectric development.

In this exercise we will use data from the National Renewable Energy Laboratory (NREL) of the US Department of Energy (DOE) to evaluate the technical and economic feasibility of expanding the use of various renewable energy sources in your locality and the United States.

Procedure

Materials Needed
Computer access to the Internet
Calculator

Directions
Wind Energy
Go to the NREL website and click on the *wind* tab. Then click on the *wind resource maps* tab. Click on the *State Wind Maps* tab that will take you to a map of the average wind speed at 80 meters above the surface. Most wind turbines are at 80 to 100 meters above the surface. You can click on your state and see the average wind speed for your location. Average wind speeds above 6.5 meters/second are considered suitable for the commercial use of wind turbines. Answer the questions on Data Sheet 14.1 under the **Wind Energy** heading.

Solar Energy
The most abundant source of renewable energy is the sun. Unfortunately it is a very diffuse source and requires technological expertise to trap it.

Go to the NREL website and click on the *PV Watts Calculator* tab. Then click on the *Grid Data Calculator (Version 2)* tab. Click on the *PVWatts Version 2 Calculator* tab. This will take you to a page where you can type in your zip code. Click *Go*. This will give you a new set of options. Select *Send to PVWatts.* This will provide you with a page that lists the coordinates of your location and specific information about the kind of solar collector being used and the price of electricity. Choose the *fixed tilt* option for the *Array Type* and input the actual cost per kWh for electricity from your utility bill in the *Cost of Electricity* position.

> **CAUTION:** Utility bills for electricity are often misleading. In addition to the actual cost of electricity consumed, there are distribution charges, fees, surcharges, taxes, and other charges. Therefore, in order to obtain the actual cost of a kilowatt-hour (kWh) of electricity it is necessary to divide the total of all costs associated with electrical use by the number of kilowatt-hours used.

Click on the *calculate* button and it will take you to a table that lists the number of kWh of electricity produced and value of the electricity produced by a 1 meter^2 photovoltaic solar collector. Answer the questions under the Solar Energy heading on Data Sheet 14.1.

Biomass Energy—Ethanol
Ethanol (ethyl alcohol) is used as a source of fuel for internal combustion engines. It is currently produced by fermenting sugar or starch, which is converted to sugar. Ethanol can be blended with gasoline in a variety of percentages. A commonly available ethanol mixture is E85 fuel, which is approximately 85% ethanol and 15% gasoline. It takes about 1.42 gallons of E85 fuel to provide the same energy as 1 gallon of gasoline. Therefore, a vehicle will not go as far on a gallon of E85 fuel as on a gallon of gasoline. In fact to be cost competitive with gasoline E85 needs to cost about 72% of the cost of an equivalent amount of gasoline.

Determine the cost per gallon of regular gasoline and E85 fuel. If you have trouble finding a local E85 source go to the Internet and type in *E85 price*. There are several sites that will give the price of E85 fuel. Go to Data Sheet 14.1 under the heading Biomass Energy Ethanol and determine if E85 is cost competitive.

Biomass Energy

On the NREL home page click on the **Resource Maps & Data** tab. This will take you to a page that has a **Dynamic Maps & GIS Data** tab in the upper left corner. Click on this tab. On the next page click on the **biomass** tab. This will take you to a series of maps that give estimates of the amount of biomass available in several categories per year by county.

The categories are:
- **Crop residues**—organic material left in the field after the crop is harvested
- **Forest residues**—organic material left after logging or other forestry practices
- **Primary mill residues**—waste from sawmills and processing of wood. 98% is currently used.
- **Secondary mill residues**—waste from woodworking shops, furniture manufacturers, etc.
- **Urban wood waste**—waste from solid waste, tree trimming, construction, and demolition.
- **Methane emissions from landfills**—decomposition of organic waste releases methane.
- **Methane emissions from manure management**—decomposition of organic waste releases methane.
- **Methane emissions from wastewater treatment**—decomposition of organic waste releases methane.

Click on the maps that give data for the different categories of biomass and determine the quantities of biomass available for your county. Since the maps give a range, such as 50-100 dry tonnes (metric tons) per year, choose the midpoint between the extremes (75 dry tonnes) to put in the table. Enter the data on Table 14.1 Estimates of Biomass Available per Year on Data Sheet 14.2 and answer the questions that follow the table.

Conversion of biomass to electricity requires about a tonne (metric ton) of biomass to produce about 3500 kWh. Calculate the number of kWh that could be produced from the biomass available from each source and enter the data in Table 14.1.

Geothermal Energy

On the NREL home page click on the **Resource Maps & Data** tab. This will take you to a page that has a **Dynamic Maps & GIS Data** tab in the upper left corner. Click on this tab. On the new page, click on the **geothermal** tab. There are two maps that you can access from this page: the Resource Potential map and the Power Generation map. Answer the questions on Data Sheet 14.3 under the Geothermal Energy heading.

Economic Analysis of Renewable Energy Technologies

On the NREL website, click on the **Energy Analysis** tab. On the new page, click on the **Transparent Cost Database Open Energy Information** tab. Click on the **Launch** tab and on the next page click on the **LCOE** tab at the upper right. LCOE stands for levelized cost of energy, which is the lifetime costs associated with building and operating a power plant divided into the electricity generated during its lifetime. You will see a graph of various electricity producing technologies and the cost of producing a kWh of electricity by this method. Since prices for technological components, fossil fuels, labor, and other factors change with time, this analysis will vary from year to year. However, the graph is updated periodically to give a reasonably accurate current measure of the cost of producing electricity from each of these technologies. If you place the mouse over the bar on the graph it will give you the median cost of producing energy from each technology.

On Data Sheet 14.4 answer the questions under the Economic Analysis of Renewable Energy Technologies heading.

EXERCISE 14
EVALUATING RENEWABLE ENERGY SOURCES

Name:_____
Section:_____
Date:_____

Data Sheet 14.1

Wind Energy
Use the state wind map to answer the following questions.
1. Is your locality suitable for the development of wind power?

2. Are there areas in your state that are suitable for the development of wind power?

3. List the best locations for wind energy development in your state.

4. If there are suitable areas that are not likely to be developed, what are the reasons for not developing wind energy at these sites?

Solar Energy
Use the solar energy calculation of the NREL website to answer the following questions.
1. How many kWh of electricity would be produced in a year by a $1m^2$ photovoltaic solar panel in your locality?

2. How many dollars would this electricity be worth?

Biomass Energy—Ethanol
1. Price of regular gasoline_____

   Price of E85 fuel_____

2. Price of regular gasoline $\times 0.72$ = _____If this number is less than the price of E85 fuel, it is more expensive to use E85 than regular gasoline.

EXERCISE 14
EVALUATING RENEWABLE ENERGY SOURCES

Name:_____
Section:_____
Date:_____

Data Sheet 14.2

Biomass Energy

Table 14.1 Estimates of Biomass Available per Year		
Category of Biomass	**Metric Tons/Year**	**Number of kWh Possible**
Crop residues		
Forest residues		
Primary mill residues	Zero, 98% already used	Zero
Secondary mill residues		
Urban wood waste		
Methane emissions from landfills		
Methane emissions from manure management		
Methane emissions from wastewater treatment		
Total		

1. A typical home uses about 11,000 kWh of electricity per year.
   How many homes could be supplied with electricity by using the biomass available in your county?

   _____

2. Determine the number of homes in the county by going to the US Census Bureau website.
   What percent of the homes in your county could be supplied with electricity from biomass?

   _____

EXERCISE 14
EVALUATING RENEWABLE ENERGY SOURCES

Name:_____

Section:_____

Date:_____

Data Sheet 14.3

Biomass Energy (continued)

3. List the biomass sources that would be the easiest to develop. Hint: Think about transportation costs, ease of collecting the resource, and the infrastructure that would need to be constructed to exploit the resource.

    a.

    b.

    c.

    d.

    Explain why you think they would be the easiest to develop.

4. What are some of the negative effects of using crop residues for energy production?

5. What are some problems associated with the use of forest residues as a source of biomass fuels?

Geothermal Energy

1. In the United States, where are the most favorable sites for the development of geothermal energy?

2. A typical fossil fuel or nuclear power plant has a generating capacity of 500–1000 megawatts. The 2010 installed capacity of the U.S. was about 1 million megawatts. If you look at the geothermal power generation map, about what percentage of the electricity could the additional planned geothermal capacity supply?

EXERCISE 14
EVALUATING RENEWABLE ENERGY SOURCES

Name:_____
Section:_____
Date:_____

Data Sheet 14.4

Economic Analysis of Renewable Energy Technologies

1. Compared to the several kinds of fossil fuel power or nuclear power plants listed, which of the renewable technologies currently are able to produce electricity for about the same cost per kWh?

2. Of the cost competitive technologies listed in the previous question, which have the greatest potential to be developed?

3. If you click on the ***Overnight capital cost*** tab you can get an idea of the cost of building a power plant per kW of power generating capacity. Which renewable energy sources are competitive with fossil fuel or nuclear power plants?

4. What effect would a reduction in the cost of solar panels have on the overnight capital cost?

5. What effect would a government subsidy have on the overnight capital cost?

EXERCISE 15
TOXICITY TESTING (LD$_{50}$)

Purpose & Objectives

The purpose of this exercise is to determine the LD$_{50}$ (concentration at which 50% of the test animals die) for several kinds of common household materials.

After completing this exercise the student will be able to:

1. Measure the effect of various toxic materials on brine shrimp (*Artemia salina*).
2. Determine the LD$_{50}$ (Lethal Dose 50%) for a variety of toxic materials.

Introduction

Every day we handle toxic materials (gasoline, oil, paint, pesticides, prescription drugs, bleach, etc.). Because many people do not appreciate the toxic nature of commonly used materials, toxic materials find their way into groundwater and waterways when they are used or disposed of improperly. Applying weed killers, insecticides, or fertilizer to lawns and fields just before a rain often results in these materials being washed into drains and streams. Often people pour unwanted toxic materials down drains or flush them down the toilet. In other cases, these materials are simply poured on the ground where they are washed into waterways or contaminate groundwater. In addition, improperly stored toxic materials enter waterways and groundwater when storage containers leak, packages are broken, or they are left exposed to the weather. When toxins enter waterways and groundwater, they contaminate drinking water sources and affect aquatic life. Aquatic organisms may be killed or weakened by exposure to toxins or their bodies may accumulate toxic molecules. In many instances, there are advisories against eating certain fish because their bodies are contaminated with toxins.

Toxins can have a wide variety of effects on an animal's biological functions. Some toxins affect important physiological functions such as respiration and have an immediate effect. Others affect functions at particular times in an organism's life, while others affect the genetic foundation of an organism. Teratogens are toxins that cause abnormalities in the development of organisms. For example, in humans, excessive consumption of alcohol during pregnancy causes mental and physical changes in the embryo. Mutagens are toxins that cause changes in the DNA of organisms, which can lead to a wide variety of problems. Carcinogens are toxins that cause disruption to the regulation of tissue growth, leading to cancer.

Determining the toxicity of materials requires a standard method of comparison. A typical method is to determine the concentration of a toxic material that cause 50% mortality in a population of test animals. This is called an LD$_{50}$ (Lethal Dose 50%) test of toxicity. For obvious reasons, we do not purposely do LD$_{50}$ studies on humans, but typically use rats or mice in toxicity studies and assume that the results we see in rats or mice are similar to what would happen in humans. However, for a variety of reasons, different species of animals respond differently to the same toxin. For example, rats are very sensitive to dioxin while humans are not. Table 15.1 lists several common materials and the LD$_{50}$ concentrations determined in rats.

Table 15.1 LD$_{50}$ of Some Common Chemicals Tested in Rats

Chemical	LD$_{50}$	Percent of body weight	
Vitamin C	11,900 mg/kg	1.2%	mg/kg is equivalent to parts per million (ppm).
Ethanol	7,060 mg/kg	0.7%	
Table Salt	3,000 mg/kg	0.3%	
Arsenic	763 mg/kg	0.08%	
Coumarin (rat poison)	293 mg/kg	0.03%	
Aspirin	200 mg/kg	0.02%	
Caffeine	192 mg/kg	0.02%	
Nicotine	50 mg/kg	0.005%	
Dioxin	20 µg/kg	0.00000002%	µg/kg is equivalent to parts per billion (ppb).
Dioxin (human)	Very low toxicity		

Procedures

Many household items that we deal with on a regular basis are toxic materials, but we don't usually think of them as being toxic. In this exercise we will determine the LD$_{50}$ concentration for solutions of copper sulfate (CuSO$_4$) and several common household chemicals to the brine shrimp, *Artemia salina*. Your instructor may want to divide the class into groups so that more kinds of materials can be tested.

Materials Needed per Group
Large quantity of newly hatched brine shrimp
Copper sulfate solution (CuSO$_4$)
Other chemical solutions: peroxide, bleach, vinegar, coffee, nicotine, alcohol, etc.
Eye droppers
Fifteen petri dishes
Tape or wax pencils for labeling petri dishes
Graduated cylinder
Magnifying glass or dissecting microscopes

Directions
Toxicity of CuSO$_4$
1. Obtain five petri dishes.
2. Label the dishes as follows:
    - 10% CuSO$_4$
    - 1% CuSO$_4$
    - 0.1% CuSO$_4$
    - 0.01% CuSO$_4$
    - 0% CuSO$_4$ (control)

(*Note: The 0% control must NOT be distilled water. Distilled water will kill the brine shrimp. Use the saltwater medium that the brine shrimp normally live in as the control.*)

3. Pour enough of each solution into the petri dish to cover the bottom of the dish.

4. Use an eyedropper to place ten brine shrimp in each of the dishes. Count the brine shrimp carefully and be sure that you do not get any unhatched brine shrimp eggs in the petri dish. (The use of a dissecting microscope or magnifying glass may be helpful.) If there are unhatched eggs, they may hatch during the experiment and make it difficult to determine an accurate LD$_{50}$.
5. Record the date and the time on the petri dishes.

Toxicity of Other Household Chemicals

1. Determine the LD$_{50}$ for two other common household chemicals. Choose from the following list, or select others with the help of your instructor. Possible choices are vinegar, rubbing alcohol, cold coffee, nicotine, and hydrogen peroxide.
   (*Do not use distilled water or tap water to mix the solutions. You will need to use the saltwater solution in which the brine shrimp normally live for mixing the solutions.*)
   i)   Take the stock solution of the household chemical you are to use as a concentration of 100%.
   ii) Prepare a 10% solution by taking 10 mL of the stock solution and adding 90 mL of brine shrimp medium to it.
   iii) To prepare a 1% solution, take 10 mL of the 10% solution and add 90 mL of brine shrimp medium.
   iv) In similar fashion, prepare a 0.1% solution and a 0.01% solution.
   v)   As a control, use the unaltered brine shrimp medium.

Data Collection

1. Examine the petri dishes of your three experiments at 24 hours and again at 48 hours and in each case record the number of brine shrimp that have died. Record your data on Table 15.2 CuSO4, Table 15.3 Household Substance 1, and Table 15.4 Household Substance 2 on Data Sheet 15.1
2. Plot your data on Graph 15.1 LD$_{50}$ CuSO$_4$, Graph 15.2 LD$_{50}$ Household Substance 1, and Graph 15.3 LD$_{50}$ Household Substance 2 on Data Sheets 15.2 and 15.3.
   i)   You will have two lines on each of the graphs: one for the number of brine shrimp that died by 24 hours and a second for the number of brine shrimp that died by 48 hours.
   ii) You may wish to enter the data into Excel and use the scatter plot option to generate the graph. You will also need to convert the concentration axis to a log scale. Because you cannot take the log of zero, enter 0.00001 for the concentration of the control.
3. Examine the graphs and determine the concentration at which 50% of the brine shrimp died at the end of 24 hours and at the end of 48 hours for each of the three experiments. You will have two LD$_{50}$ concentrations: one for 24 hours and one for 48 hours.
   i)   For example, in the hypothetical graph shown below, the dark line represents the mortality at 48 hours and crosses 5 (half the organisms died) between concentration 0.01 and 0.1 at approximately a 0.05% concentration.
   ii) The lighter colored line represents mortality at 24 hours and crosses 5 between 0.1 and 1 at approximately a 0.3% concentration.

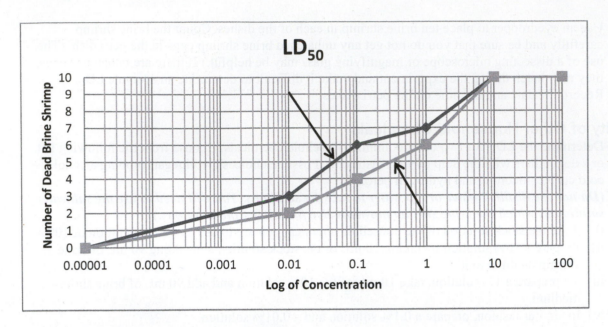

Figure 15.1 Example Graph of Data

EXERCISE 15
TOXICITY TESTING (LD$_{50}$)

Name:_____
Section:_____
Date:_____

Data Sheet 15.1

Table 15.2 LD$_{50}$ 24 and 48 hours

CuSO$_4$ Concentration	Number of Dead Brine Shrimp	
	24 hours	48 hours
100%		
10%		
1%		
0.1%		
0.01%		
0% (Control)		

Table 15.3 LD$_{50}$ 24 and 48 hours

Household Substance 1 _____ Concentration	Number of Dead Brine Shrimp	
	24 hours	48 hours
100%		
10%		
1%		
0.1%		
0.01%		
0% (Control)		

Table 15.4 LD$_{50}$ 24 and 48 hours

Household Substance 2 _____ Concentration	Number of Dead Brine Shrimp	
	24 hours	48 hours
100%		
10%		
1%		
0.1%		
0.01%		
0% (Control)		

EXERCISE 15
TOXICITY TESTING (LD$_{50}$)

Name:_____

Section:_____

Date:_____

Data Sheet 15.2

Graph 15.1 LD$_{50}$ CuSO$_4$

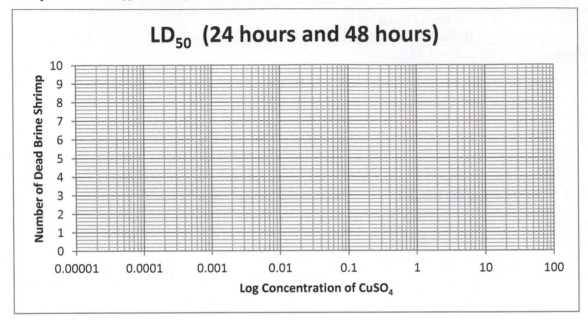

Graph 15.2 LD$_{50}$ Household Substance 1_____

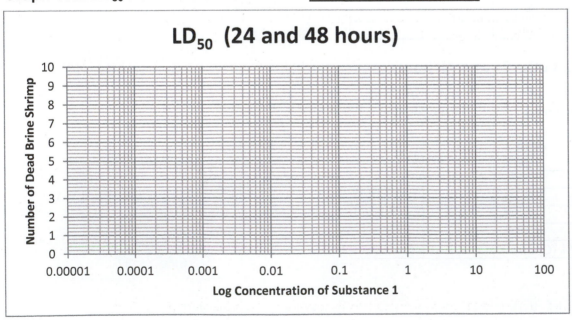

EXERCISE 15
TOXICITY TESTING (LD$_{50}$)

Name:_____
Section:_____
Date:_____

Data Sheet 15.3

Graph 15.3 LD$_{50}$ Household Substance 2_____

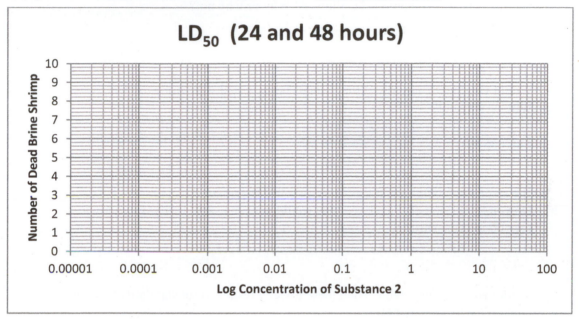

Analysis

1. A control is an important part of any experiment.
    a. What is the purpose of the control in each experiment?

    b. Did all the brine shrimp survive in all the controls?

    c. Was there a difference in the number of brine shrimp that died in the controls at 24 hours and 48 hours?

    d. What might have contributed to any difference in the number that died at 24 and 48 hours?

2. Which of the substances you tested was the most toxic to brine shrimp?

EXERCISE 15
TOXICITY TESTING (LD$_{50}$)

Name:_____

Section:_____

Date:_____

Data Sheet 15.4

3. LD$_{50}$ tests have been called inhumane because test animals die during the test.
    a. Should LD$_{50}$ determinations be allowed? Why or why not?

    b. Would you feel differently if the animals being tested were mice, rabbits, or monkeys? Why or why not?

4. How do the results you obtained in this exercise relate to humans?
    a. Do you think each of the compounds would have the same level of toxicity in humans as they do in brine shrimp?

    b. Would you be comfortable using these toxicity results to set standards for exposure for humans? Why or why not?

5. The level at which you first detect an effect is called the threshold level. Was there a concentration below which you could not detect an effect on the number of brine shrimp that died in the CuSO$_4$ solutions or the solutions of other household chemicals?

EXERCISE 15
TOXICITY TESTING (LD$_{50}$)

Name:_____
Section:_____
Date:_____

Data Sheet 15.5

6. Would an LD$_{50}$ toxicity test be useful in evaluating teratogens, mutagens, or carcinogens? Why or why not?

7. Obtain the Material Safety Data Sheets (MSDS) for the chemicals you tested. Simply go to the Internet and type in MSDS followed by the name of the chemical.
   a. What are the published LD$_{50}$s for the substances you tested?

   b. How do the LD$_{50}$s from the MSDS compare to the LD$_{50}$s you determined for brine shrimp?

EXERCISE 16
EFFECTS OF SALINIZATION ON PLANTS

Purpose & Objectives

The purpose of this exercise is to examine the effects of salinization on plants. After completing this exercise, the student will be able to:

1. Identify and describe the physical effects on plants of growing in saline soils.
2. Discuss why plants exposed to excess salt in soils experience physiological stress.
3. Analyze and predict trends in plants exposed to the effects of increasing soil salinity.
4. Recommend mitigation and management strategies for areas at risk of salinization.

Introduction

Salinization is the accumulation of salts in the soil due to human activities. The use of salts on roadways to prevent ice from forming is a common practice that can increase soil salinity. Excess salts from roads are carried to nearby soils by runoff during heavy rain events and as snow and ice accumulations melt. Common irrigation practices can also lead to an accumulation of salts in the soil. According to the USDA, almost 50% of irrigated land world-wide has seen a reduction in crop productivity due to salinization. Areas most at risk for salinization are agricultural fields in arid climates that make extensive use of irrigation to supplement low rainfall that also experience high evaporation rates. Areas reporting some of the highest percentage of saline-affected soils are Australia, Africa, and the Middle East.

Salts are compounds formed by the combination of positively charged ions (cations) and negatively charged ions (anions). Most commonly, salt compounds in soils form from several cations such as sodium (Na^+), magnesium (Mg^{2+}), calcium (Ca^{2+}), and potassium (K^+) and a few anions such as chloride (Cl^-), bicarbonate (HCO_3^-), and sulfate (SO_4^{2-}). Irrigation water naturally contains soluble salts that dissolve in water as it travels through soil and rock. Since irrigation is used in dry climates, some of the water applied to fields evaporates leaving salt behind in the soil. Over time, salt can accumulate and cause reduced crop yields. Many plant species are sensitive to saline soils and when exposed to excess salts they show signs of water stress, such as leaf burn/discoloration and stunting of growth. This is because increased salt in the soil reduces the roots' ability to take in water. Also, high levels of salinity can alter soil composition and affect soil chemistry.

There are several agricultural management practices that can be implemented to decrease the risk of salinization. These practices include but are not limited to:

- Utilizing sustainable practices for irrigation such as drip irrigation systems, soil moisture monitoring, and center-pivot irrigation to reduce the amount of irrigation water needed.
- Providing better drainage systems in agricultural fields.
- Implementing cropping and tillage practices that optimize water infiltration and soil permeability.
- Planting of salt-tolerant plants either as main crop or as cover crops.

Since salts dissolved in water allow water to conduct electricity, a standard way to measure salinity is to use a conductivity meter. A conductivity meter determines how well current flows through a solution and gives a reading in decisiemens per meter (dS/m).

Procedure

Materials Needed per Group of 2–3 Students

Sodium chloride (30–40g)

Plant Seeds [Note: You should choose plant species that grow quickly such as mung beans (*Vigna ridiata*), radish (*Raphanus sativus*), or Wisconsin Fast Plants (*Brassica rapa*)

Planting containers 3-4 inches (7-10 cm) across (8)

Potting soil mix that contains nutrients

Grow light

Conductivity meter and/or probe

Deionized or distilled water (4L)

1000 mL Beakers (4)

Electronic balance

Glass stirring rods (3)

Labeling tape and marker and/or wax pencil

Initial Setup

1. Obtain **8** planting containers and fill each container ¾ of the way full with potting mix. Label two containers each as follows: control, slightly saline, moderately saline, and highly saline.

2. The group is responsible for determining what they will use as a control as well as the saline solution concentrations they will make to represent the following treatment groups: slightly saline, moderately saline, and highly saline. [Note: your saline solutions should fall within the range of 3dS/m to 26 dS/m.]

3. Select the dS/m concentration for the group's control and three treatment groups that fall within the range 3dS/m to 26 dS/m. Record the four dS/m concentrations in in the appropriate columns on Table 16.1 on Data Sheet 16.1.

4. Create the different saline solutions by adding the appropriate amount of sodium chloride (NaCl). For every 1dS/m you should add 0.67g/L. For example, if you were to make a 4dS/m saline solution you should add 2.68g (4 ∗ 0.67) of NaCl to 1 liter of deionized water.

5. Use the electronic balance to accurately determine the grams of NaCl needed to make your first solution based on the dS/m concentration your group selected.

6. Add the NaCl to 1L of deionized water in a beaker. Stir vigorously for 3 minutes or until all the salt is visibly dissolved. You may wish to check the accuracy of your saline solution with a conductivity probe (measures in dS/m). Label the beaker containing the solution with its treatment designation (control, slightly saline, moderately saline, or highly saline) and its dS/m determination.

7. Repeat steps 4 through 6 to make the remaining solutions.

8. If using mung beans or radish seeds, place two seeds approximately 3cm apart in your container. Push them down into the soil to a depth of 1.0 to 2.0 cm and cover with soil. If using Wisconsin Fast plants follow the guidelines on their website www.fastplants.org.

9. For each plant container, add the appropriate saline solution until the soil is "damp"— it should not be completely saturated. Continue to monitor soil moisture during the length of your investigation (1 to 2 weeks). If the soil is "dry to the touch" add the appropriate solution to the soil until it is damp. This may require daily visits to the laboratory or your instructor may make provisions to make sure the plants are kept watered appropriately. You may need to make more saline solution depending on how long you run the investigation.

10. Place each container under a grow light.

11. Each group will make one measure of the effect of salinity, which will involve measuring the height of plants. In addition your group will also decide on two additional quantitative (can observe and measure) plant characteristics on which to collect data that you think will be related to the effects of soil salinization on plants. You should also determine how often you need to collect these data during the length of the investigation.

Data Collection

12. At the end of the investigation, each group will determine the average plant height for each of the four sets of plants and record the data on Table 16.1 on Data Sheet 16.1. Also record plant height data for all class groups on Table 16.2 on Data Sheet 16.1. If more than one group collected data for a specific salinity (dS/m), a new average including all of the plants grown under that specific salinity will need to be calculated. [Note: There may not be data recorded for each dS/m concentration, since some concentrations may not have been tested by any of the groups in the class.]

13. Record the information pertaining to the two additional quantitative characteristics on Table 16.3. Be sure to provide an appropriate title for the table.

14. On Data Sheet 16.3, create Graph 16.1 that shows average plant height for various salinity solutions as recorded in Table 16.2.

15. On Data Sheet 16.4 and 16.5, create Graphs 16.2 and 16.3 that shows the data results of the additional quantitative studies recorded in Table 16.3. Be sure to provide an appropriate title and y-axis label for the graph.

EXERCISE 16
EFFECTS OF SALINIZATION ON PLANTS

Name:_____
Section:_____
Date:_____

Data Sheet 16.1

Table 16.1 Effect of Salinity on Plant Height Measurements

	Control	Slightly Saline	Moderately Saline	Highly Saline
Concentration (dS/m)				
Average Plant Height (cm)				

Table 16.2 Class Plant Height Measurements

Concentration (dS/m)	Average Plant Height (cm)	Concentration (dS/m)	Average Plant Height (cm)
0		14	
1		15	
2		16	
3		17	
4		18	
5		19	
6		20	
7		21	
8		22	
9		23	
10		24	
11		25	
12		26	
13			

EXERCISE 16
EFFECTS OF SALINIZATION ON PLANTS

Name:_____
Section:_____
Date:_____

Data Sheet 16.2

Table 16.3 _____

	Control	Slightly Saline	Moderately Saline	Highly Saline
Concentration (dS/m)				

EXERCISE 16
EFFECTS OF SALINIZATION ON PLANTS

Name:_____

Section:_____

Date:_____

Data Sheet 16.3

Analysis

Graph 16.1 Effect of Salinization on Plant Height (Total Class Data)

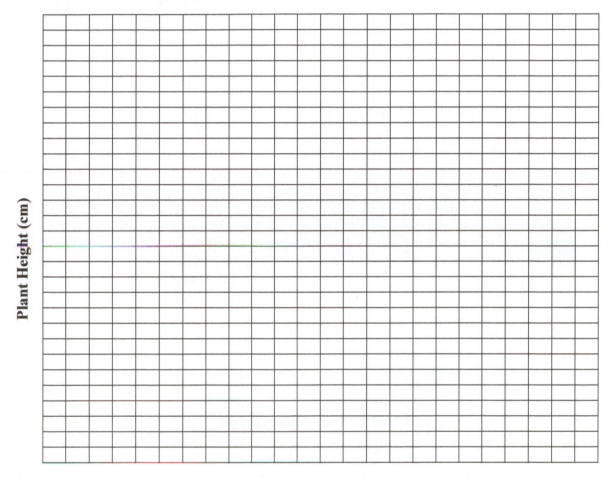

EXERCISE 16
EFFECTS OF SALINIZATION ON PLANTS

Name:_____
Section:_____
Date:_____

Data Sheet 16.4

Graph 16.2 _____

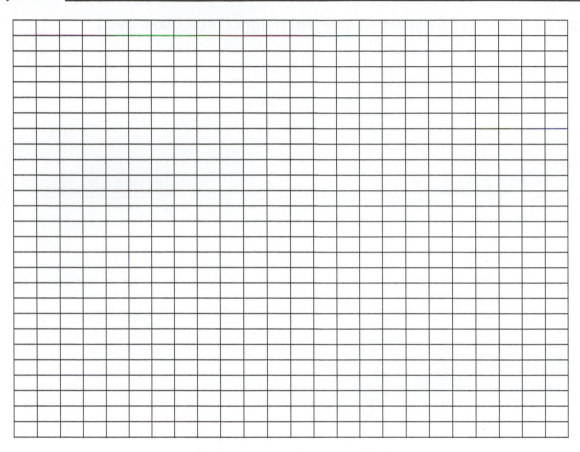

Saline Concentration (dS/m)

EXERCISE 16
EFFECTS OF SALINIZATION ON PLANTS

Name:_____
Section:_____
Date:_____

Data Sheet 16.5

Graph 16.3 _____

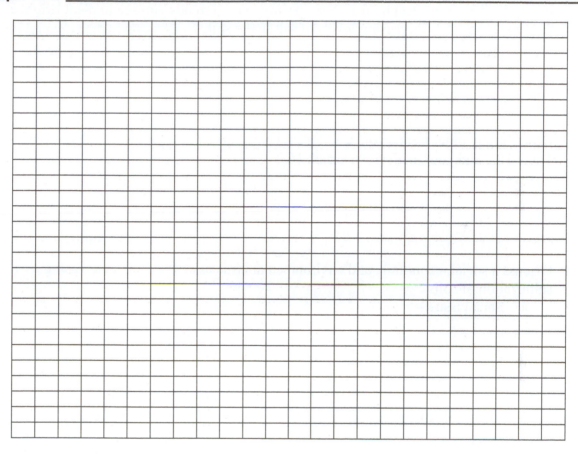

Saline Concentration (dS/m)

EXERCISE 16
EFFECTS OF SALINIZATION ON PLANTS

Name:_____

Section:_____

Date:_____

Data Sheet 16.5

Questions

1. Using the data from Graph 16.1, describe any overall trends you see in plant height as they are exposed to increasing levels of salt in the soil.

2. Describe any other trends you see in plant characteristics based on your data from Table 16.3 and your analysis in Graph 16.2.

3. Discuss a social, economic, and environmental issue associated with increasing salinization of soils.
    a. Social issue –

    b. Economic issue –

    c. Environmental issue –

4. The Central Valley of California is a highly productive agricultural area that is dependent on irrigation water from both surface and groundwater resources. Recently, an apricot farmer in the Central Valley has noticed white crust appearing on the top soil of her farm. She has also seen an overall stunting of growth in the plants and discoloration in the leaves of the trees.
    a. Why is this area at greater risk for salinization of her agricultural land?

    b. Explain why the plants are experiencing stunted growth and discoloration.

c. Describe two management practices that can reduce the risk of salinization of soils.

EXERCISE 17
DISSOLVED OXYGEN AND BIOCHEMICAL OXYGEN DEMAND

Purpose & Objectives

The amount of oxygen dissolved in water is critical to the survival of aquatic organisms. Therefore, oxygen levels are often used as an indicator of the health of a body of water. In this exercise we will measure the amount of oxygen present in water, determine the percent saturation, and determine the biochemical oxygen demand of water samples. After completing the work associated with this exercise, the student will be able to:

1. Perform a test to determine the dissolved oxygen content of a water sample.

2. Use a chart to determine the percent saturation of oxygen in a water sample.

3. Determine the biochemical oxygen demand (BOD) of a water sample.

Introduction

The amount of oxygen dissolved in water is important because oxygen is necessary for respiration by aquatic organisms. Dissolved oxygen (DO) in water is measured in milligrams/liter or parts of oxygen per million parts of a solution (ppm). (mg/L and ppm are essentially the same.)

Different organisms require different amounts of dissolved oxygen in order to survive. Trout, for example, need roughly 6.5 ppm, carp need 2.5 ppm, and sludge worms can live in 0 ppm of dissolved oxygen. The amount of dissolved oxygen in water is influenced by several factors such as the temperature of the water, turbulence, the amount of suspended and dissolved organic matter, respiration by aquatic organisms, and photosynthesis by aquatic plants and microorganisms. The presence of a high oxygen level in water is an indication that the water quality is good, while a low oxygen content indicates poor quality water.

Human activities can be a major factor contributing to changes in dissolved oxygen levels in water. Organic wastes from human sewage, concentrated animal feeding operations, and industries are an important problem because decomposer bacteria use oxygen from the water to metabolize organic compounds. The runoff of fertilizer from farms and lawns is also a major problem. The nutrients in fertilizer stimulate the growth of aquatic plants and photosynthetic microorganisms. When the weather is cloudy or night falls, photosynthesis is reduced or stopped, but all organisms (including those that carry on photosynthesis) still continue to use oxygen for respiration and oxygen levels fall. Furthermore, when these large populations of plants and microorganisms die, aerobic bacteria consume oxygen in the process of decomposing the dead organic matter and oxygen levels fall.

The presence of organic matter (either natural or human caused) results in the growth of decomposer organisms that use oxygen for aerobic respiration as they break down organic matter. This demand for oxygen is known as the **biochemical oxygen demand (BOD).** A common method for determining the effect of organic matter on the oxygen content of water involves comparing the oxygen concentration of two water samples. The first sample is taken directly from the water source and the amount of dissolved oxygen is determined. A second sample is taken at the same time but is sealed and kept in the dark at 20°C for 5 days. The decomposer organisms in the water of the second sample continue to break down organic matter in the water and use up oxygen in the process. After 5 days the second sample is tested for its dissolved oxygen content. The difference between the initial sample and the second sample is the biochemical oxygen demand.

(DO sample 1) – (DO sample 2) = BOD

Downstream from a point where organic material enters the water (such as a municipal sewage plant discharge), a characteristic decline and restoration of water quality can be detected either by

measuring the dissolved oxygen content or by observing the flora and fauna that live in successive sections of the river. The oxygen decline and subsequent rise downstream is called the **oxygen sag curve** (see Figure 17.1). Above the pollution source, oxygen levels support normal populations of clean-water organisms. Immediately below the source of pollution, oxygen levels begin to fall as decomposers metabolize organic materials. Rough fish, such as carp, bullheads, and gar, are able to survive in this oxygen-poor environment, where they eat both decomposer organisms and the waste itself. Farther downstream the water may become anaerobic (without oxygen), so that only the most resistant microorganisms and invertebrates can survive. Eventually most of the organic matter is consumed, decomposer populations become smaller, and the water becomes oxygenated once again as a result of diffusion of oxygen into the water and photosynthetic activity. Depending on the volumes and flow rates of the effluent plume and the river receiving it, normal communities may not appear for several miles downstream.

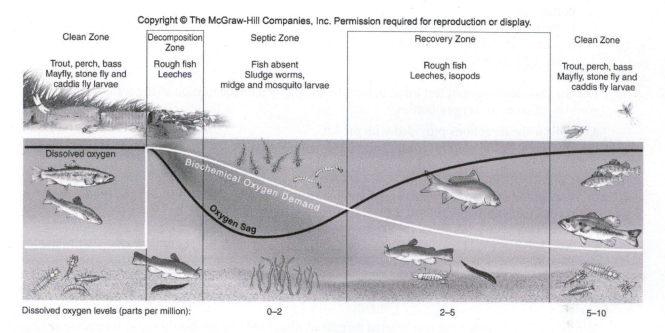

Figure 17.1 Oxygen Sag Curve

Procedure

Materials Needed per Group
Thermometer

Dissolve oxygen test kit

Four 300 mL glass stoppered dissolved oxygen bottles (They should be black with stoppers that provide an airtight seal.)

A one liter bottle of distilled water to rinse containers used during the dissolved oxygen test

A plastic jar, with a screw on lid, to be used to collect any wastes produced by the oxygen test

Equipment Needed in Lab
Incubator

Activity (First Class Meeting)

In this exercise we will evaluate the water quality of two different bodies of water. Your instructor will divide the class into groups and direct each group to specific bodies of water or give suggestions for sources you may visit. At each of the two sites you will do the following:

1. Identify the site on Table 17.1 on Data Sheet 17.1.

2. Take the temperature of the water. Record the temperature on Table 17.1 on Data Sheet 17.1.

3. Collect 2 water samples. This should be done carefully so that you get a good sample.
    a. Take the sample away from the bank or shore.
    b. Collect the sample such that you do not stir up sediments that enter the dissolved oxygen bottle.
    c. Submerge the bottle and allow it to fill. Place the glass stopper into the top of the bottle while it is still submerged so that you do not trap any air bubbles.
    d. Label both bottles with the name of the site, the date, and your name.

4. Use the dissolved oxygen test kit to determine the oxygen concentration of the water from one of your two dissolved oxygen bottles.
    a. Follow the directions provided with the kit.
    b. Place any liquid or solid waste from the test into the waste collection jar.
    c. Record the dissolved oxygen content you obtain on Table 17.2 Initial Dissolved Oxygen Content on Data Sheet 17.1.
    d. Repeat the oxygen test on a second sample from your initial dissolved oxygen bottle, and record the result on Table 17.2 on Data Sheet 17.1.
    e. If the two results are **NOT** the same, repeat the test a third time.
    f. Average the **TWO** results that are nearly the same and enter the result on Table 17.1 in the Dissolve Oxygen (Initial) column on Data Sheet 17.1.

5. Repeat steps 1-4 at the second site.

6. Return to the lab. Place the two labeled stoppered dissolved oxygen bottles in an incubator at 20°C for 5 days. (It would be permissible to leave them for 7 days if it fits schedules better.)

Activity (Second Class Meeting)

1. Use Figure 17.2 to determine the % saturation of the initial water samples.
    a. Place a straight edge so that it connects the temperature of the water from the intial sample with the oxygen concentration you determined.
    b. Where the straight edge crosses the diagonal line you can read off the percent saturation.
    c. Record the results on Table 17.1 in the Percent Saturation Column.

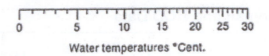

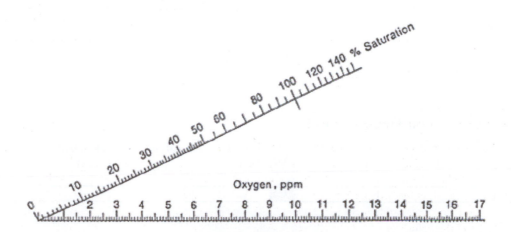

Figure 17.2 Percent Saturation Graphic

2. Identify your two dissolved oxygen bottles in the incubator and take them to your lab station.

3. Choose one of the bottles, note the source of the sample, and perform the dissolved oxygen test on the water in the bottle, two times.
    a. Record your results on Table 17.2 in the Five Days Dissolved Oxygen column.
    b. If the two results are **NOT** the same, repeat the test a third time.
    c. Average the **TWO** results that are nearly the same and enter the result on Table 17.1 in the Dissolve Oxygen (After 5 days) column on Data Sheet 17.1.

4. Repeat the procedure on your second dissolved oxygen water sample.

5. Determine the BOD for your two sites.
    a. Subtract the dissolved oxygen concentration of the 5-day sample from the dissolved oxygen concentration of the initial sample. For example, if the initial DO concentration was 8 mg/L and the oxygen level after five days was 3 mg/L, then the BOD would be 5 mg/L. (8 mg/L – 3 mg/L = 5 mg/L)

    b. Enter your results on Table 17.1 in the BOD column.

EXERCISE 17
DISSOLVED OXYGEN AND BIOCHEMICAL
OXYGEN DEMAND

Name:_____

Section:_____

Date:_____

Data Sheet 17.1

Table 17.1 Dissolved Oxygen and BOD Data

Source of Sample	Water Temperature (°C)	% Saturation	Dissolved Oxygen (Initial)	Dissolved Oxygen (After 5 days)	BOD

Table 17.2 Dissolved Oxygen Tests

Site	Initial Dissolve Oxygen Content				Five-Day Dissolved Oxygen Content		
	Test 1	Test 2	Test 3		Test 1	Test 2	Test 3
1							
2							

Analysis

1. The following factors affect the oxygen concentration in water.

Temperature	Higher temperature gives lower DO
Salinity	Higher salinity gives lower DO
Photosynthesis	Higher DO
Respiration	Lower DO
Barometric Pressure	Higher pressure gives higher DO
Altitude	Higher altitude gives lower DO
Diffusion	Balances O_2 concentration between air and water

a. If you were to take your samples just before sunrise, how might your results be different? Why?

b. Compare a rapidly flowing mountain stream with a small, slow-moving river. Which is likely to have a high dissolved oxygen concentration? Why?

EXERCISE 17
DISSOLVED OXYGEN AND BIOCHEMICAL
OXYGEN DEMAND

Name:_____
Section:_____
Date:_____

Data Sheet 17.2

2. Why was it important to keep the second sample in the dark in a stoppered bottle in order to determine BOD?

3. If you measured the temperature of a stream as 2°C and it had an oxygen concentration of 8 mg/L would it be a healthy stream? Why or why not?

4. Which of the two test sites had the highest BOD rating?
    What factors did you observe at the site that might have contributed to the high BOD?

5. Describe 2 reasons the BOD of an unpolluted lake or stream is likely to be higher in the fall than in the spring.

6. A typical BOD for an unpolluted water source is about 5 mg/L. Based on this standard are your sites unpolluted, moderately polluted, or highly polluted with organic matter?

EXERCISE 18
AIR POLLUTION

Purpose & Objectives

In this exercise we will examine the various levels of gases emitted from vehicles of differing efficiency. Student will also record ambient levels of both ozone and temperature. After completing this exercise, students will be able to:

1. Measure concentration of gases released from motor vehicles—including CO_2, CO, NO, and SO_2.
2. Examine ozone levels and evaluate, based on the EPA's Air Quality Index, the level of health concern for humans.
3. Analyze how combustion efficiency in motor vehicles contributes to the emission levels of primary air pollutants in the troposphere.

Introduction

As you know, our air is comprised of a mixture of many different molecules. Approximately, 99% of our air is made up of nitrogen (78%) and oxygen (21%). The remaining 1% is largely comprised of argon and carbon dioxide.

An air pollutant is any substance, in high enough concentrations, that can cause environmental damage or human health effects. The severity of air pollution in a given area depends on several factors, including climate, topography, population density, and the number and type of industrial activities. Pollutants typically take two basic forms – gases such as carbon monoxide (CO), sulfur dioxide (SO_2), volatile organic compounds (VOCs), and oxides of nitrogen (NO, N_2O, NO_2) and particulate matter such as pollen, smoke, dust, and fly ash. Most ambient air pollutants are directly released into the troposphere, the layer of the atmosphere near the Earth's surface, in an unmodified form and are known as primary pollutants. Primary pollutants come from a variety of sources that include both natural and anthropogenic activities. These sources are classified as either mobile (motor vehicles) or stationary (industrial smokestack, volcanoes). Once in the troposphere, and in the presence of sunlight, which catalyzes chemical reactions, they often mix with one another or with naturally occurring molecules to form secondary pollutants. A common secondary pollutant is tropospheric ozone (O_3), a major component in photochemical smog, which forms from oxides of nitrogen combining with VOCs through catalytic processes driven by sunlight.

Due to the detrimental human health impacts and environmental damage of air pollution, the Clean Air Act was enacted, which requires that the Environmental Protection Agency set standards, known as the National Ambient Air Quality Standards, for six common air pollutants in the United States (see Table 18.1). The EPA has also established an Air Quality Index that calculates the potential health risk for humans based on the daily concentrations of ground-level ozone, particulate matter, carbon monoxide, and sulfur dioxide (see Table 18.2). In 2007, the U.S. Supreme Court ruled that the EPA has the authority, under the Clean Air Act, to regulate heat-trapping gases such as carbon dioxide. Currently, the EPA does not set regulations on carbon dioxide emission, but it does monitor carbon dioxide concentrations as a part of the Inventory of U.S. Greenhouse Gas Emissions and Sinks report that began in 1990. The EPA monitors the concentration of all greenhouse gases and analyzes trends in greenhouse gas concentrations to better understand the potential impact of climate change. The EPA may decide in the future to set regulations for carbon dioxide emissions, especially from coal burning power plants.

Table 18.1 EPA's Six Principal Air Pollutants

Pollutant	Source(s)	Human Health Impact	Environmental Impact
Carbon Monoxide	Motor vehicles (incomplete combustion)	Reduces hemoglobin's ability to carry oxygen – can cause death	Short-lived pollutant often converted to CO_2
Sulfur Dioxide	Coal burning power plants	Respiratory irritation; aggravates asthmatic conditions	Primary component involved in acid deposition
Oxides of Nitrogen	Motor vehicles (NO); agricultural practices (N_2O)	Respiratory irritation; aggravates asthmatic conditions	NO and NO_2 are involved in acid deposition and photochemical smog formation
Lead	Industrial source – metal smelting	Can cause neurological disorders	Reduction in biodiversity for organisms sensitive to lead contamination
Ozone	Forms as a secondary pollutant from NOx + VOC	Respiratory irritation; aggravates asthma and emphysema	Reduces plant growth and increases susceptibility to disease
Particulate Matter	Smokestacks, fires, construction sites, farming activities	Affects lung and heart health; in extreme cases can lead to cancer and death	Can increase acid deposition effects in ecosystems, reduces visibility

Table 18.2 EPA's Air Quality Index

AQI Values	Levels of Health Concern	Ground-Level Ozone Concentration
0 – 50	Good	0 – 59 ppb
51 – 100	Moderate	61 – 75 ppb
101 – 150	Unhealthy for sensitive groups	76 – 95 ppb
151 – 200	Unhealthy	96 – 115 ppb
201 – 300	Very unhealthy	116 – 374 ppb

Materials
Materials Needed per Group
- Gastec precision gas analysis apparatus (GV 50PS or GV 100S)
- Printed copy of sampling guide
  - GV 50PS http://www.gastec.co.jp/english/reference/c8kids.htm
  - GV 100S http://www.gastec.co.jp/english/products/seihin/c9_pdf/en/GV-100S_GV-110S.pdf
- (3) Carbon monoxide gas detector tubes
- (3) Carbon dioxide gas detector tubes (0.5% – 8% range)
- (3) Nitric oxide gas detector tubes
- (3) Sulfur dioxide gas detector tubes
- PVC pipe
- (1) Ozone test strips
- Outdoor thermometer

Procedures
1. Students should be placed in groups of four or five, if possible.
2. Each group should select three motor vehicles from which they will sample the exhaust. Students should conduct some background information on their vehicles to ensure they represent a range of efficiency—such as a hybrid, SUV, and a motorcycle. Students may be able to find the vehicle efficiency on the manufacturer's website or on http://www.fueleconomy.gov/.
3. Record your vehicle's MPG and any fuel considerations—such as use of ethanol, diesel, hybrid engine, electric vehicle— in Table 18.3 on Data Sheet 18.1.
4. Once all three vehicles are selected, attach the PVC pipe to the end of the tail pipe of the first vehicle.
5. Start the vehicle and allow it to run for 2–3 minutes prior to taking samples.
6. Follow the appropriate procedures as outlined in the Gastec user's manual for taking samples of the exhaust.
7. One student should start the vehicle as the other students in the group immediately take samples of carbon dioxide, oxygen, and nitric oxide emissions. Students should stand to the side of the PVC pipe, not directly behind it. Be sure to tilt the tip of the syringe into the end of the PVC pipe so that the sample is taken directly in the airflow coming out of the vehicle's tail pipe.
8. Record the sampling results in Table 18.4 on Data Sheet 18.1.
9. Repeat procedures 3 through 7 for the next two vehicles.
10. Use an ozone test strip product to determine the ozone concentration (ppb) of the air. Also determine the air temperature and time of day. Record ozone concentration, air temperature, and time in Table 18.5 on Data Sheet 18.1.
11. Create a bar graph of your emission sampling results on Graph 18.1 on Data Sheet 18.2. You may choose to record your data per vehicle or per pollutant. You may also want to use Excel to create your graphs. For any graph you create, be sure to include appropriate x and y axes titles and a key to identify the three different vehicles sampled.

EXERCISE **18**
AIR POLLUTION

Name:_____

Section:_____

Date:_____

Data Sheet 18.1

Table 18.3 MPG of Motor Vehicles Sampled

	Vehicle 1: _____	Vehicle 2: _____	Vehicle 3: _____
MPG			
Fuel Type			

Table 18.4 Concentration of Motor Vehicle Emissions

Gas Sample	Vehicle 1: _____	Vehicle 2: _____	Vehicle 3: _____
Carbon Monoxide			
Carbon Dioxide			
Nitric Oxide			
Sulfur Dioxide			

Table 18.5 Ozone Concentration and Ambient Air Temperature

Ozone Concentration	Ambient Outdoor Temperature	Time of Day

EXERCISE 18
AIR POLLUTION

Name:_____
Section:_____
Date:_____

Data Sheet 18.2

Graph 18.1 Gas Emissions Comparisons for Three Motor Vehicles

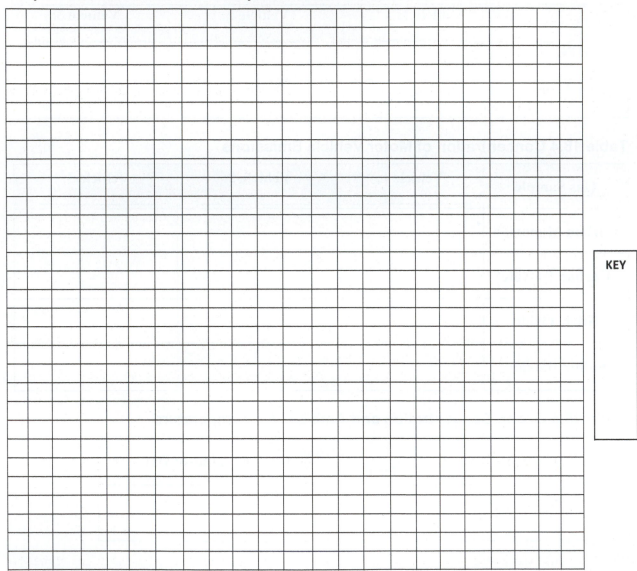

KEY

EXERCISE 18
AIR POLLUTION

Name:_____
Section:_____
Date:_____

Data Sheet 18.3

Analysis

1. Compare the trends from your graphical analysis of pollutant emissions with each vehicle's MPG rating.
    a. Do you see any overall correlation between MPG and emission of carbon dioxide and nitric oxide?

    b. If yes, explain what accounts for any correlations you see. If no, explain why there are no correlations.

2. Describe how the level of sulfur dioxide compares to the other gases you sampled in this exercise. Explain what accounts for this difference.

3. Answer the questions below regarding Air Quality Index. You may want to refer to the chapter on ground-level ozone in your textbook prior to answering the questions below.
    a. What AQI index did your ozone sample measurement fall into? What is the health recommendation for this level?

    b. Why was it important to record the time and temperature for this measurement?

4. In this exercise we measured carbon dioxide concentration, even though this is not one of the six pollutants regulated by the EPA. Discuss why this is still an important gas to monitor and therefore why it was included in this investigation.

5. Based on what you learned from this lab evaluate what steps you could take to:
    a. Reduce your overall emissions of carbon monoxide, carbon dioxide, and nitric oxide from your transportation habits.

    b. Reduce the level of ground-level ozone that is present in your city.

EXERCISE 19
GLOBAL INDICATORS OF CLIMATE CHANGE

Purpose & Objectives

This exercise will utilize data sets from the Environmental Protection Agency (EPA), National Aeronautics and Space Administration (NASA), and National Oceanic and Atmospheric Administration (NOAA) to examine global indicators of climate change. After completing this exercise, students will be able to:

1. Identify the major greenhouse gases and explain how they contribute to global climate change.
2. Analyze graphical representations of changes to global surface temperature, sea level, and land and sea ice concentrations and draw conclusions about trends.
3. Evaluate changes in their personal habits that could reduce greenhouse gas emissions.

Introduction

The **weather** at a particular location includes short-term (days to weeks) measurements of things such as temperature, wind, humidity, atmospheric pressure, sunshine, cloudiness, and precipitation. A region's **climate** is an average of weather conditions over many years and includes seasonal changes in weather patterns throughout the year (summer—winter, rainy—dry, etc.). The weather and climate of a region are largely determined by the geographical location of the region as well as its proximity to large bodies of water.

In addition to local and regional weather and climate, it is possible to look at average conditions for the entire Earth (global climate). One of the unique features of Earth's climate compared to most other planets is its relatively high temperature. The temperature is determined by a balance between the amount of energy it receives from the sun and the amount of energy lost to space. Approximately 26% of the incoming solar radiation never makes it to Earth's surface as it is reflected by the atmosphere and cloud cover. Roughly 4% of the incoming solar radiation is reflected from Earth's surface. The proportion of solar radiation reflected from Earth's surface is known as **albedo**. Surfaces differ in their ability to reflect solar radiation. For example, white surfaces such as snow have a higher albedo than dark surfaces such as soil. Much of the incoming solar radiation is absorbed by Earth's surface and its atmosphere and is radiated back to space in the form of heat (infrared radiation). Gases in the atmosphere known as greenhouse gases (water, CO_2, CH_4, N_2O) absorb some of the outgoing radiation and in turn release some of this energy back towards the surface of the Earth, causing warming, known as the **greenhouse effect** (see Figure 19.1). The Earth's global mean surface temperature is approximately 14°C (57°F) but would be only approximately -18°C (0°F) if it were not for the natural greenhouse effect.

According to scientists, the Earth's average temperature has increased between 0.5 – 0.9°C (1.0 – 1.7°F), over the last 100 years. Since there is a strong correlation between increasing temperatures and increasing concentrations of greenhouse gases, scientists believe the increase in the Earth's average temperature is due to human activities that have caused an unnatural, rapid increase in the concentration of greenhouse gases in our troposphere (lower atmosphere). Numerous human activities that involve burning fossil fuels (industrial production, electrical production, and transportation), waste incineration and storage, and agricultural practices are responsible for releasing greenhouse gases to the atmosphere. According to the

EPA in 2010, carbon dioxide is the primary greenhouse gas emitted from human activities and accounted for 84% of all U.S. greenhouse gas emissions.

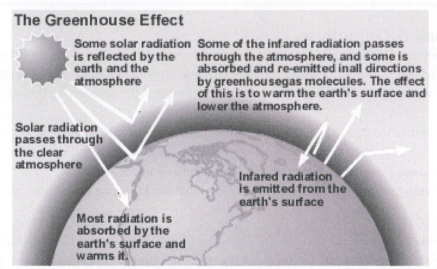

Figure 19.1 The Greenhouse Effect (Courtesy of the U.S. EPA)

Scientists are currently recording and monitoring many global indicators of climate change. Some of the primary changes being monitored are

Sea-Level Rise – Sea-level rise is due primarily to thermal expansion (water molecules get farther apart) of the ocean as it warms in response to increasing global temperatures. The melting of land ice, such as glaciers, is also contributing to sea-level rise by adding more water to the oceans.

1) Melting of Arctic Sea Ice – Increasing temperatures in both the atmosphere and the oceans is causing a reduction in the extent and thickness of sea ice.

2) Changing Weather Patterns – As the atmosphere warms it will shift current precipitation patterns causing droughts in some areas and increased precipitation in others. Scientists also predict that there will be more frequent and severe storms and hurricanes.

3) Ocean Acidification – Increasing concentrations of CO_2 in the atmosphere result in more CO_2 dissolving into the ocean increasing the acidity in some regions.

Materials

Calculators

Rulers

Internet Connection

Procedures

Part A

1. Use data from Tables 19.1, 19.2, and 19.3 to plot changes in carbon dioxide, nitrous oxide, and methane emissions from 2005 to 2010 on Graph 19.2 on Data Sheet 19.1. (Data for the tables is taken from the EPA's *Inventory on United States Greenhouse Gas Emissions and Sinks: 1990 – 2010).*

2. Use Graph 19.1, World Carbon Dioxide Emissions, Tables 19.1 and 19.2, and your Graph 19.2 on Data Sheet 19.1 to answer the questions in Part A Greenhouse Gas Emission on Data Sheets 19.2 and 19.3.

Parts B–F

1. Go to NASA's homepage on climate located at http://climate.nasa.gov/

2. On the left-hand tool bar, click on the section titled "key indicators."

3. From this section you will be utilizing background information, graphical data, and interactive maps to explore global indicators of climate change and make decisions on how to reduce your greenhouse gas footprint.

   a. Carbon Dioxide Concentration: Read the directions and answer the questions on Data Sheet 19.3 Part B.

   b. Global Surface Temperature: Read the directions and answer the questions on Data Sheet 19.4 Part C.

   c. Arctic Sea Ice and Land Ice: Read the directions and answer the questions on Data Sheet 19.4 Part D.

   d. Sea Level: Read the directions and answer the questions on Data Sheet 19.5 Part E.

   e. Reducing Your Greenhouse Footprint:

      (i) First calculate your carbon footprint on the EPA's website - http://epa.gov/climatechange/ghgemissions/individual.html. Read the introduction to this activity on the website before clicking on the carbon calculator in the middle of the page. Answer the questions about your carbon footprint once you finish the activity and record your results in Part F on Data Sheets 19.5 and 19.6.

      (ii) Answer the corresponding questions about reducing your contribution to methane and nitrous oxide emissions in Part F on Data Sheet 19.6.

Graph 19.1 World Carbon Dioxide (CO_2) Emissions

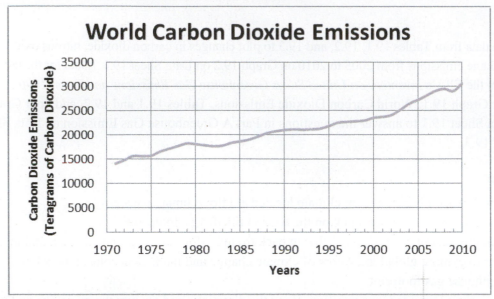

Data from the International Energy Agency

Table 19.1 EPA Inventory of US Greenhouse Gas Emissions and Sinks: 1990–2010

(Tg or million metric tons CO_2 Eq)

Gas / Source	1990	2005	2006	2007	2008	2009	2010
CO₂ Total*	**5,100.5**	**6,107.6**	**6,010.0**	**6,118.6**	**5,924.3**	**5,500.5**	**5,706.4**
Fossil Fuel Combustion	4,738.3	5,746.5	5,653.0	5,757.8	5,571.5	5.206,2	5,387.8
Electricity	1,820.8	2,402.1	2,346.4	2,412.8	2,360.9	2,146.4	2,258.4
Transportation	1,485.9	1,896.6	1,878.1	1,893.9	1,789.8	1,727.9	1,745.5
Industrial	846.4	816.4	848.1	844.4	806.5	726.6	777.8
Waste Incineration	8.0	12.5	12.5	12.7	11.9	11.7	12.1

*The entire data set was not included and therefore the sources given will not add to the total
emission of carbon dioxide.

Table 19.2 EPA Inventory of US Greenhouse Gas Emissions and Sinks: 1990–2010

(Tg or million metric tons CO_2 Eq – The phrase **CO_2 Eq** means that the methane concentrations have been adjusted to be comparable in effect to equivalent carbon dioxide concentrations)

Gas / Source	1990	2005	2006	2007	2008	2009	2010
CH_4 Total*	668.3	625.8	664.6	656.2	667.9	672.2	666.5
Natural Gas Systems	189.6	190.5	217.7	205.3	212.7	220.9	215.4
Enteric Fermentation	133.8	139.0	141.4	143.8	143.4	142.6	141.3
Landfills	147.7	112.7	111.7	111.7	113.1	111.2	107.8
Coal Mining	84.1	56.8	58.1	57.8	66.9	70.1	72.6
Manure Mgt	31.7	47.9	48.4	52.7	51.8	50.7	52.0
Rice Cultivation	7.1	6.8	5.9	6.2	7.2	7.3	8.6

*The entire data set was not included and therefore the sources given will not add to the total emission of methane.

Table 19.3 EPA Inventory of US Greenhouse Gas Emissions and Sinks: 1990–2010

(Tg or million metric tons CO_2 Eq – The phrase **CO_2 Eq** means that the nitrous oxide concentrations have been adjusted to be comparable in effect to equivalent carbon dioxide concentrations)

Gas / Source	1990	2005	2006	2007	2008	2009	2010
N_2O Total*	316.2	331.9	336.8	334.9	317.1	304.0	306.2
Agricultural Soil Management	200.0	213.1	211.1	211.1	212.9	207.3	207.8
Stationary Combustion	12.3	20.6	20.8	21.2	21.1	20.7	22.6
Mobile Combustion	43.9	37.0	33.7	29.0	25.2	22.5	20.6
Manure Management	14.8	17.6	18.4	18.5	18.3	18.2	18.3
Nitric Acid Production	17.6	16.4	16.1	19.2	16.4	14.5	16.7

*The entire data set was not included and therefore the sources given will not add to the total emission of nitrous oxide.

EXERCISE 19
GLOBAL INDICATORS OF CLIMATE CHANGE

Name:_____
Section:_____
Date:_____

Data Sheet 19.1

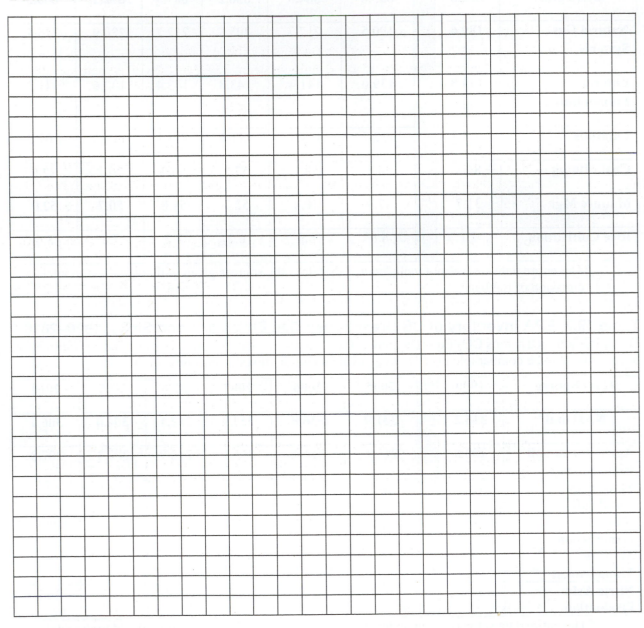

EXERCISE 19
GLOBAL INDICATORS OF CLIMATE CHANGE

Name:_____
Section:_____
Date:_____

Data Sheet 19.2

Part A. Greenhouse Gas Emissions

1. Use Graph 19.1 of world carbon dioxide emissions and Table 9.1 of U.S. carbon dioxide emissions and calculate the percent of world emissions contributed by the United States in 1990 and 2010.

    1990: _____   2010: _____

2. Use Graph 19.1 and Table 19.1 for CO_2 emission to answer the following questions:

    a. Calculate the percent change in U.S. carbon dioxide emission from 1990 to 2010 (Note: Percent change is calculated by subtracting the 1990 figure from the 2010 figure and dividing by 1990 figure).

        _____

    b. Now calculate the percent change in **global** carbon dioxide emissions from 1990 to 2010.

        _____

    c. Describe some factors that could account for the difference in global CO_2 emission rate of increase versus the U.S. CO_2 emissions rate of increase.

        _____
        _____
        _____

3. Use Table 19.2 for CH_4 emission to answer the following questions:

    a. Calculate the percent change in methane emissions from 2005 to 2010.

        _____

    b. Enteric fermentation is a process in ruminant organisms, such as cows, that converts a percentage of the food digested into CH_4. Describe changes in human activities that would lead to an increase in enteric fermentation emissions in cattle.

        _____
        _____
        _____

    c. Most rice is grown in flooded fields called paddies. Describe how this method of rice cultivation contributes to methane emissions.

        _____
        _____
        _____

    d. How do natural gas production, landfills, coal mining, and animal manure management activities contribute to methane production?

        _____
        _____

EXERCISE 19
GLOBAL INDICATORS OF CLIMATE CHANGE

Name:_____
Section:_____
Date:_____

Data Sheet 19.3

4. Use Table 19.3 for N_2O emission to answer the following questions:
   a. Describe how agricultural soil management could contribute to nitrous oxide emissions.

   _____
   _____

   b. Provide an example of both a stationary source and a mobile source of nitrous oxide emissions.

   _____
   _____
   _____

   c. Explain why mobile emission sources have seen a significant reduction since 1990.

   _____
   _____
   _____

Part B. Carbon Dioxide Concentration

Read the information provided in the carbon dioxide concentration section of the NASA website. Next, use both the indirect and direct measurement graphs of global carbon dioxide concentration to answer the following questions.

1. Prior to 1950, what was the highest recorded concentration of carbon dioxide on record?

   _____

2. How do scientists record carbon dioxide concentrations indirectly?

   _____

3. Based on direct measures, what was the % increase of carbon dioxide concentration from 2005 to 2012? [Note: if you hover your computer mouse arrow over the line on the graph it will provide specific data points for each year.]

   _____

4. The graph showing the direct measurement for carbon dioxide concentrations in the troposphere from 2005 to 2012 has been corrected (averaged) for season variations. Explain why global carbon dioxide concentrations would fluctuate with the Earth's seasons.

   _____
   _____

5. Identify the season of the Northern Hemisphere that would have the highest CO_2 concentrations and explain why this is true.

   _____
   _____

EXERCISE 19
GLOBAL INDICATORS OF CLIMATE CHANGE

Name:_____
Section:_____
Date:_____

Data Sheet 19.4

Part C: Global Surface Temperature

Read the information provided in the global surface temperature section of the NASA website. Next, use the Global Land-Ocean Temperature Index graph to answer the following questions:

1. The global surface temperature graph shows the deviation from the long-term average for the period 1951–1980. The temperature anomaly is the degrees above or below the average. If you compare the mean temperature of 1920 with the mean temperature in 2010, how large a difference is there in temperature in °C?

   _____

2. Study the Global Surface Temperature graph from 1880 to 2010 and write a descriptive qualitative statement that describes the overall trend during this period.

   _____
   _____

Use the Time Series: 1884–2010 interactive map to answer the following questions:

3. In 1920, where was the largest warming anomaly found? Approximately how many degrees above the long-term average was the anomaly?

   _____

4. In 2010, where was the largest warming anomaly found? Approximately how many degrees above the long-term average was the anomaly?

   _____

5. Write a statement that describes the overall trend in global surface temperature measurements from 1884 to 2010. Hint: Move the red indicator from 1884 to 2010 and note the change in color.

   _____
   _____

Part D: Arctic Sea Ice and Land Ice Loss

Read the information provided in arctic sea ice and the land ice sections of the NASA website and then answer the following questions.

1. If arctic sea ice extent is declining by 11.5% every decade (based on 2012 estimates), use the data recorded in 2010 (in million square km) to predict what arctic sea ice extent will be in 2020.

   _____

EXERCISE 19	Name:
GLOBAL INDICATORS OF CLIMATE CHANGE	Section:
	Date:

Data Sheet 19.5

Part D: Arctic Sea Ice and Land Ice Loss (Continued)

2. As arctic and land ice decreases in surface coverage at the poles, this area will experience a decrease in albedo (reflection rate of incoming solar energy). Explain how this can cause a positive feedback loop.

    _____

    _____

3. The Antarctic ice mass is losing about 100 cubic kilometers of ice per year. Over 20 years this amounts to 2000 cubic kilometers of ice. If this volume of ice were converted to an equivalent volume of water and converted to a layer 1 cm thick how many km^2 would it cover? Hint: If you were to cut 1 cm layers out of the km^3, how many 1 cm layers could you get out of a km^3. The surface area of the oceans of the world is approximately 352,000,000 km^2. What effect would the melting of 2000 km^3 of Antarctic ice have on sea level?

    _____

    _____

    _____

Part E: Sea Level

Read the information provided in sea level section of the NASA website. Next, use the information in this section and the Ground Data and Satellite Data graphs to answer the following questions:

1. What two consequences of a warmer climate are thought to be the cause of sea level rising?

    a. _____

    b. _____

2. a. Calculate the rate of change in sea level (in mm) between 1870 and 1930.

        _____

    b. Now calculate the rate of change in sea level (in mm) between 1930 and 1990.

        _____

    c. What could account for such an increase in the rate of change after the 1930s?

        _____

EXERCISE 19
GLOBAL INDICATORS OF CLIMATE CHANGE

Name:_____
Section:_____
Date:_____

Data Sheet 19.6

Part E: Sea Level (Continued)

3. Use the current rate of change from satellite data to calculate what the sea level increase could be by 2100, assuming the rate of change remains the same.

    _____

Part F: Reducing Your Greenhouse Gas Footprint

Use the carbon calculator from the following website to determine your greenhouse gas footprint.

http://epa.gov/climatechange/ghgemissions/individual.html

1. What was your carbon footprint, according to the EPA carbon calculator?

    _____ lbs of CO_2 / year

2. How does your average compare to the typical U.S. average for the same size household?

    _____

3. Describe three things you indicated you could do to reduce your overall carbon footprint.

    a. _____

    b. _____

    c. _____

4. Review the sources of methane in Table 19.2. Identify and describe two things you could do to reduce your overall contribution to methane emissions.

    a. _____

    b. _____

5. Review the sources of nitrous oxide in Table 19.3. Identify and describe two things you could do to reduce your overall contribution to methane emissions.

    a. _____

    b. _____

EXERCISE 20
EVALUATING ECOLOGICAL FOOTPRINT CALCULATIONS

Purpose & Objectives

There are several organizations that provide ecological footprint calculators on the internet. You will choose two of these and answer the questions that are part of the calculation. Finally, you will explore why the questions are relevant. After completing this exercise the student will be able to:
1. Compare the results of two different ecological footprint calculators.
2. Identify ways students could reduce their ecological footprint.
3. State why certain lifestyle factors are important to determining an ecological footprint.

Introduction

The **ecological footprint** is a measure of human demand on Earth's renewable resources. It is typically represented by the amount of land and water resources needed to provide the things we use and to absorb our wastes. Renewable resources can continue to provide services indefinitely if the resources are used sustainably. However, there is a maximum rate at which renewable resources, such as fisheries, agricultural land, forests, air quality, freshwater, and renewable energy sources can provide services. Currently, the pressure of human use on Earth's renewable resources is equivalent to about 1.5 Earths. Obviously, that is not sustainable.

Furthermore, the impact of the "footprint" may actually be felt in a different part of the world from the demand behavior that puts pressure on renewable resources. For example, if your lifestyle requires coffee, the impact of your demand for coffee is transferred to the land used by farmers in tropical countries where coffee can be grown. If demand increases and farmers attempt to meet that demand, they may switch to unsustainable farming practices that lead to planting on unsuitable land, increased use of pesticides, or increased use of fertilizer that leads to water pollution.

There is a second aspect of the concept of an ecological footprint that recognizes that some individuals use much more than their fair share of the Earth's resources, while others are able to use only a small fraction of their fair share of Earth' resources. The lifestyle of the developed countries of the world is typically not sustainable but is in a sense subsidized by the lower living standards of people in the less-developed world.

Many organizations maintain websites that allow you to calculate your ecological footprint. Typically, the result of such an exercise is to produce a statement about how many Earths it would take to satisfy your lifestyle. The basic idea behind ecological footprint calculations is that resources are limited and some people have access to and use many more resources than others. These sites typically ask a series of questions about your lifestyle, which are then used to determine your ecological footprint. Obviously there are certain assumptions built into the program. For example, there will be questions about transportation, housing, food habits, and purchasing behavior.

Procedure

1. Connect to the internet and use the search phrase "ecological footprint."

2. Choose two different sites that allow you to calculate your personal ecological footprint.

3. Complete the two calculations and record your data on Table 20.1 on Data Sheet 20.1.

4. Compare the two sites you visited by listing strengths and weaknesses on Table 20.2 on Data Sheet 20.1.

5. Choose the one site that you felt was the most thorough and list 5 lifestyle changes you would be able to make and record these on Table 20.3 on Data Sheet 20.1.

6. Repeat the calculation on the website you considered to be most thorough but modify your responses to match the lifestyle changes you entered on Table 20.3. Enter the new ecological footprint calculation in the Size of Ecological Footprint with Modified Lifestyle column of Table 20.1.

EXERCISE 20
EVALUATING ECOLOGICAL FOOTPRINT
CALCULATIONS

Name:_____

Section:_____

Date:_____

Data Sheet 20.1

Table 20.1 Ecological Footprint

Website Name	Size of Ecological Footprint	Size of Ecological Footprint with Modified Lifestyle

Table 20.2 Comparison of Two Ecological Footprint Websites

	Website Name_____	Website Name_____
Strengths	1.	1.
	2.	2.
	3.	3.
	4.	4.
	5.	5.
Weaknesses	1.	1.
	2.	2.
	3.	3.
	4.	4.
	5.	5.

Table 20.3 Lifestyle Changes You Could Make

1.
2.
3.
4.
5.

EXERCISE 20
EVALUATING ECOLOGICAL FOOTPRINT
CALCULATIONS

Name:_____
Section:_____
Date:_____

Data Sheet 20.2

Analysis

1. How effective were the lifestyle changes you selected at reducing your ecological footprint?

2. List 3 reasons why food habits are important to the calculation of an ecological footprint.
    a.

    b.

    c.

3. List 2 reasons why transportation was important to the calculation of an ecological footprint.
    a.

    b.

4. Why are the kind of housing and the number of people living in a house important to determining an ecological footprint?

5. Why are buying habits important to determining an ecological footprint?

EXERCISE 21
LAND-USE PLANNING ON CAMPUS

Purpose & Objectives

College campuses range from those in a rural setting with much open space to urban, multistory buildings in the city center. Regardless of the setting, it is possible to design a campus so that it makes wise use of the space it occupies, is aesthetically attractive, and provides open space.

After completing this exercise, the student will be able to:

1. Identify sources of information necessary to conduct a land-use assessment of a campus.
2. Use a checklist to inventory the quality of land-use planning on a campus.
3. Develop a list of proposals that could improve aspects of land-use planning on campus.

Introduction

Good land-use planning involves evaluating many aspects of a site, before it is modified, and then producing a design for the buildings and landscape that fulfills the objectives of the project. Most college campuses were established many years ago, so some of the original plans envisioned by the founders may have been lost. In addition, as campuses have grown, new thinking about the appropriate use of space has changed. In this exercise you will use a check list of important land-use considerations to evaluate how well your campus meets common land-use planning principles. This will involve evaluating the original plan and more recent adjustments to the original plan. This may involve meeting with the college officials involved in planning and those involved in maintaining the buildings and grounds.

There are many things to be considered whenever a land-use decision is made. It is helpful to have some guidelines to consider when evaluating a land-use change. Although the following list is not exhaustive, it provides some helpful ideas about things that should be considered.

1. Evaluate and record any unique geologic, geographic, or biologic features of the land.
2. Preserve unique cultural or historic features.
3. Conserve open space and environmental features.
4. Recognize and calculate the cost of additional infrastructure changes that will be required to accommodate altered land use.
5. Plan for mixed housing and commercial uses of land in proximity to one another.
6. Plan for a variety of transportation options.
7. Set limits and require managed growth with compact development patterns.
8. Encourage development within areas that already have a supportive infrastructure so that duplication of resources is not needed.

Procedure

In this exercise you will use a variety of resources to explore how land-use planning principles are used on your campus.

> *In order to obtain the resources you need, you must interact with campus and local officials. When dealing with busy people it is important that you come to an interview prepared, so that you use their time efficiently. Have your list of questions or topics preselected and if you are a part of a group have a designated spokesperson for the group.*

1. Organize the class into work groups with assigned responsibilities.

2. Obtain a map of the college campus. It should be as complete and accurate as possible.

3. Also use Google Earth to obtain an aerial view of the campus. This will give you a better idea of vegetation than a campus map, which is designed to help people find places on campus.

4. Contact the local government planning body. This could be a city, county, township, or regional agency. If available, obtain a land-use planning map and any documents that state the planning principles used by the planning agency.

5. Make an appointment with the campus director of the building and grounds department. Select questions from the checklist that specifically relate to the care of the grounds and buildings. If necessary it would be permissible to do the interview over the phone.

6. Determine who is responsible for land-use planning decisions on campus and make an appointment with that person. Select questions from the checklist that relate directly to long-range planning. If necessary it would be permissible to do the interview over the phone.

7. Use the checklist of land-use topics to assess land use on your campus. Actually walk the campus while doing the assessment. On large campuses you may want to organize into groups that will survey specific parts of the campus.

8. Following a class discussion prepare a list of proposals for ways in which land use on campus could be improved. List your consensus on Table 21.1 on Data Sheet 21.1.
   In preparing your list consider the following:

Rationale—	Why is the change desirable?
	What land-use principles does it support?
Practicality—	Is the change easy to implement?
	Are people invested in maintaining current practices rather than welcoming change?
Economics—	Would the change cost or save money?
	Are there long-term savings after a significant up-front cost?
External forces—	What forces outside the campus could support or hinder the change?
	Are there grants that would support the change?
	Are there powerful interests that would support or challenge the change?
	Are there local, state, or federal ordinances or laws that bear on the proposed change?

Checklist of Land-Use Topics

Site history
>   When was the campus built?
>   What was present on the site before the campus was built?

Unique geologic, geographic, or biologic features of the land
>   List features you feel are worthy of protection.
>   What unique features have already been preserved?
>   Are there features that should have been preserved but were not?

Unique cultural or historical features of the site
>   What cultural or historical features have already been preserved?
>   Are there cultural or historical feature that should have been preserved but were not?

Open space and environmental features
>   Describe examples of good use of open space.
>   Describe examples of poor use of open space.

Water management issues
>   Are there wetlands, streams, ponds, or floodplains that were included in the land-use plan?
>   Were there any wetlands or ponds that were drained?
>   Are any streams free-flowing or have they been dammed, straightened, or enclosed in concrete?
>   Does the vegetation used in landscaping match the availability of water?
>   Is there evidence that the bodies of water are consciously included in the land-use plan?
>   Is there evidence that the existence of a flood plain was taken into account in the planning process?
>   How does run-off water leave campus?
>   How does run-off water leave parking lots?

Landscape management/biodiversity
>   What percentage of the land surface of the campus is lawn?
>   Are irrigation systems or lawn sprinklers used to maintain lawns?
>   What pest management practices are used?
>   Are there places where the natural vegetation (woodlots, prairie, desert, etc.) survives and is protected?
>   What percentage of the plantings (flowers, trees, and shrubs) are native species?
>   Animals need food, water, and shelter. What actions are used to encourage wildlife populations?
>   Do any wildlife populations cause problems (damage to landscape plants, safety hazards, disease, etc.)?

Transportation issues
>   Is it easy to walk from one building to another?
>   Are there paths and parking areas for bicycles?
>   What percentage of the land surface is devoted to parking and roads?
>   Are there bus or train stations adjacent to campus?
>   Are there multistory parking structures?

Energy issues

What percentage of the student body drives to campus?

Are there incentives for students to carpool?

List examples of buildings that have been remodeled to improve energy efficiency.

Air quality issues

Are there places on campus that have unpleasant odors?

Is dust a problem?

Solid waste planning issues

How is waste from the management of the grounds managed?

What percent of campus waste is recycled, reused, or composted?

How is food waste handled?

How is hazardous waste from laboratories, medical facilities, and machinery handled?

What steps have been taken to reduce the quantity of hazardous waste?

Current and future expansion

Have recent building projects involved new land being modified or old buildings being renovated or repurposed?

How are decisions made about the siting of new facilities?

Is there a comprehensive long-term plan?

Aesthetic issues

List the places on campus that you consider to be unappealing.

Are there ways to change these sites so that they are less offensive?

Integration of campus planning with other units of government

Is the campus planning function integrated with the local governmental planning agency?

Are campus officials members of the local governmental planning agency?

Do college officials need to request permission from the local governmental planning agency to site buildings, change the flow of streams, or other land-use practices?

EXERCISE 21
LAND-USE PLANNING ON CAMPUS

Name:_____
Section:_____
Date:_____

Data Sheet 21.1

Table 21.1 Proposals for Changes to Land Use on Campus

Proposal 1	
Rationale	
Practicality	
Economics	
External Forces	
Proposal 2	
Rationale	
Practicality	
Economics	
External Forces	
Proposal 3	
Rationale	
Practicality	
Economics	
External Forces	
Proposal 4	
Rationale	
Practicality	
Economics	
External Forces	

EXERCISE 22
SOLID WASTE ASSESSMENT

Purpose & Objectives
In this exercise we will examine the kinds and amounts of solid waste each of us produces and assess ways to reduce the amount of material needing disposal.
After completing this exercise the student will be able to:
1. Describe the kinds and amounts of solid waste they personally produce.
2. Determine the proportion of the waste that could be recycled.
3. State ways in which individuals can reduce the amount of solid waste they produce.

Introduction
On average each person in North America produces about 2 kilograms (4.4 pounds) of solid waste daily. Over a year this amounts to 730 kilograms (1600 pounds). On average about 34 percent of solid waste is recycled. The volume of solid waste produced is related to two primary social factors: personal lifestyle decisions and costs associated with the management and disposal of waste. Lifestyle factors include such personal decisions as reducing the number of food and beverage containers purchased, avoiding unnecessary packaging, purchasing items that will last a long time rather that those that need frequent replacement, and repairing items rather than replacing them.

The cost of disposal is a strong economic incentive to people to make changes that reduce the amount of waste produced. Some areas have high disposal costs because of a lack of adequate sites for landfills or a reliance on more expensive incineration systems. This cost is passed on to the public in the form of taxes or fees for municipal solid waste services.

In 2010 the U.S. EPA published the following information (Table 22.1) about the average composition of solid waste. Although waste types such as yard waste vary widely in different regions, these averages help to characterize the nature of solid waste.

Table 22.1 Municipal Solid Waste Generation

Type of waste	Percent of total
Paper and Paperboard	28.5%
Food Scraps	13.9%
Yard Trimmings	13.4%
Plastics	12.4%
Metals	9.0%
Rubber, Leather, and Textiles	8.4%
Wood	6.4%
Glass	4.6%
Other	3.4%

Procedures

In this exercise you will compile records of all the solid waste you dispose of over a three-day period.

1. Over the course of at least three days, collect all the items you dispose of. Separate your solid waste into the following categories.
   a. Paper
   b. Glass
   c. Plastic
   d. Metal
   e. Food waste
   f. Other wastes

2. For items such as paper, glass, plastic, metal, and other waste you simply may want to place each category of solid waste in a separate plastic bag or other container.
   a. At the end of the three days sort each kind of waste (paper, plastic, glass, etc.) into recycled and non-recycled items. Separately weigh the quantity of recycled and non-recycled items in each category and record your results on Data Sheets 22.1 and 22.2.
   b. Then, further subdivide the items into the categories listed in the tables (containers, packaging, other).
   c. Inventory the items in each category. For example, in the plastic category you may have some items that are recycled and some that are not. So on the Not Recycled table on Data Sheet 22.1 you might list 3 yogurt containers, 1 catsup bottle, and 1 piece of plastic wrap. On the Recycled table on Data Sheet 22.2 you might list 6 soft drink bottles, 2 milk jugs, and 1 salad dressing bottle.

3. Food items are a little more difficult to manage. You will need to separate food waste into solids and liquids.
   a. Use two plastic, glass, or metal sealable containers to separately store your solid and liquid food wastes.
   b. Weigh the empty containers with the lids on them.
   c. Record on the appropriate Data Sheet the amount and kind of food each time you put something into either of the two containers. For example, in the solid container you might list ½ hot dog, bone from steak, and ½ cup uneaten vegetables. In the liquid category you might list ¼ bottle of soda, ½ cup of coffee, and ¼ bowl of soup.
      Food wastes that are composted should be listed on the Recycled Data Sheet.
   d. If you weigh the empty containers before you start, you will be able to determine the weight of solid or liquid food disposed of by simply weighing the full container and subtracting the weight of the empty container and will not need to handle the food waste.

EXERCISE 22
SOLID WASTE ASSESSMENT

Name: _____
Section: _____
Date: _____

Data Sheet 22.1

Waste (not recycled)

	Paper		Plastic		Glass		Metal		Food Waste		Other Waste	Total
	Packaging	Other	Packaging	Containers	Other	Containers	Other	Containers	Other	Solids	Liquids	All items	
Total Items													
Total Weight													

167

EXERCISE 22
SOLID WASTE ASSESSMENT

Name: _____
Section: _____
Date: _____

Data Sheet 22.2

Waste (recycled)

	Paper		Plastic		Glass		Metal		Food Waste		Other Waste		Total	
	Packaging	*Other*	*Packaging*	*Containers*	*Other*	*Containers*	*Other*	*Containers*	*Other*	*Solids*	*Liquids*	*All items*		
Total Items														
Total Weight														

EXERCISE 22
SOLID WASTE ASSESSMENT

Name:_____

Section:_____

Date:_____

Data Sheet 22.3

Analysis

The U.S. EPA statistics on solid waste include that generated by business and industry as well as household waste. However, you can obtain some insight into your personal solid waste generation by comparing your daily generation with that of the nation as a whole.

1. How does your daily waste generation rate differ from the average of 2 kg/person/day?

    Why is it different?

2. On average 34 percent of solid waste is recycled. How does your recycling rate compare with the average?

    Why is it different?

3. Of the solid waste you generated, what percent of the total could have been recycled in your community?

4. What category of waste (paper, food waste, plastic, etc.) constituted the largest proportion of your waste?

    How does this compare with national averages?

5. If you were to engage in a program to reduce the amount of waste you produce, what three categories would be easiest for you to reduce?

    Why?

APPENDIX 1
THE SUCCESSFUL FIELD TRIP

Introduction

The purpose of a field trip is to provide hands-on experience that helps students understand how theory and practice are interrelated. Every field trip should have a clearly established set of goals and objectives. These should be discussed and written down to focus the participants' attention on the important parts of the experience, so that they are not distracted by sights or activities that are peripheral to the trip's purpose. The participants may need to prepare for the trip by reading background material and discussing the roles each will have in meeting the objectives of the trip experience.

Procedures

1. Discuss and establish a clear set of objectives for the class to meet during the trip. Write these down and provide each student with a copy.
2. Assign responsibilities to class members.
3. Develop an equipment list if you plan to do an exercise that will require equipment.
4. Do a dry run if possible, so that you can work the bugs out of the procedure before you go on the trip.
5. Develop a report form, data sheet, or checklist, so that students can record their progress in meeting the objectives of the trip.
6. Plan a time for discussion and analysis of the data collected during the field trip, the errors that were made, and how effectively the learning experience met the planned objectives. If different groups had different assignments, they will need to report these results to the rest of the class. If different groups were collecting the same kind of data, have them pool their data with the data from other groups for analysis.
7. If the trip involves a visit to a place that has not been used previously, the instructor should visit the site to become familiar with the geography, hazards, and important features of the site.
8. Transportation needs to be arranged.
9. For legal liability reasons, the school administration will require that certain practices and procedures are followed. Find out what these are and file the appropriate paperwork.
10. If the field trip involves outdoor fieldwork, be sure to provide students with information about proper clothing, and to be prepared for changes in the weather.
11. Each student should have a specific assignment for the day. If the class is divided into small groups with specific assignments, it is less likely that anyone will be left out.
12. Establish a clear set of acceptable and unacceptable behaviors. This is especially important when taking extended trips of a day or two.
13. Use outside resource people at the site if they are available.
14. If specimens are to be collected, establish guidelines about the number of individual organisms that should be collected for identification and study in the lab or as additions to an established collection. Avoid collecting large numbers of organisms that are just going to be thrown out when you get back to the lab.

APPENDIX 2
SUGGESTIONS FOR FIELD TRIPS AND ALTERNATIVE LEARNING ACTIVITIES

Ecological Principles

Field Trip Suggestions

1. Visit local habitats—grassland, desert, forest, or other locally available habitats.
   Collect five plants that you can take back to the lab to identify.
   Collect five invertebrate animals. Record in detail where each animal was found. Upon returning to the laboratory, identify the animals.
2. Visit various aquatic systems, such as a stream, a lake, a bog, an irrigation canal, or another locally available aquatic system.
   Collect five plants that grow in or adjacent to the water that you can take back to the lab to identify.
   Collect five invertebrate animals. Record in detail where each animal was found. Upon returning to the laboratory, identify the animals.
3. Visit various controlled ecosystems, such as a sewage treatment plant, an agricultural field, a municipal park, a forestry plantation, or another locally available area. Describe five ways in which each controlled ecosystem is different from a similar natural one.
   Describe five organisms that are aided by the human control of ecosystems and five that are harmed.
4. Visit a farm and record five examples of good soil conservation practices and five examples of poor soil conservation practices.

Alternative Learning Activities

1. Track an animal in the mud or snow. Identify the animal and try to determine what it was doing from the tracks you find.
2. Place some soil in a broth made by boiling hay and observe the changes in protozoa present over a two-week period.
3. Describe the stages of succession you can observe along the edge of a pond, in an abandoned field, or in a sidewalk crack.
4. Photograph several stages of succession you can observe in your neighborhood.
5. Participate in a local habitat modification project aimed at increasing the numbers of certain kinds of organisms.
6. Dig a hole or observe a road cut and identify the different layers of a soil profile.
7. Observe the food preferences of different species of birds at bird feeders. What types of food are eaten? Does seed size make a difference? Does the kind of feeder make a difference? Do some birds only feed on the ground?
8. Dissect a recently living frog and look for parasites in the lungs, body cavity, urinary bladder, and intestines.

Population Topics

Field Trip Suggestions

1. Visit an orchard, cotton field, or vegetable farm. Determine which pest populations must be controlled and the pesticides that are used to control the pests. Figure out the cost of the pesticides used each year. Determine what special training is required for pesticide applicators. What conditions determine when pesticides are applied?

Alternative Learning Activities

1. Have a person from your local extension service office come to the class to discuss the pros and cons of pesticide use and the licensing requirements for pesticide applicators in your state.

2. Research the history of Planned Parenthood and the contribution of Margaret Sanger to the movement. Write a paper on the history of reproductive enlightenment.

3. Survey a community and ask questions about desirable family size and actual family size. Correlate your data with the age of the respondent.

4. Prepare a graph showing the population growth of your local urban community over the past twenty years. Select five changes (political, economic, sociological, housing, etc.) in the community, which appear to be related to population change and describe how they are related.

Pollution Issues

Field Trip Suggestions

1. Visit a local industry's water pollution-control facilities. Determine the cost of operating the facility. What are the pollution problems the industry is trying to control? What techniques are used?

2. Canoe down a local stream or river through an industrial area, and record examples of pollution. Consult with your local pollution-control agency about your findings. Do they agree or disagree that the examples you cite are important?

3. Fly over or drive through an industrial area, and record examples of pollution. Consult with your local pollution-control agency about your impressions. Do they agree or disagree that the examples you cite are important?

4. Visit your local water treatment plant. Make a list of the treatment methods employed and what purpose each method accomplishes. Ask what the major water-quality problems are in your area.

5. Visit your local solid waste disposal site. How many tons of waste materials are buried per day? What efforts are there to recover resources, rather than just dispose of them?

Alternative Learning Activities

1. Pick up the litter along a road or section of a beach or stream. Inventory the waste found. Identify the causes of litter and discuss solutions to the problem.

2. Ask a spokesperson from your local water planning board to talk to the class about where water is used and how water use will affect the economy of the area.

3. Invite a spokesperson from an environmental group to talk to the class about the major pollution problems in your area.

4. Collect newspaper articles that relate to pollution. Note the sources used in writing the article. If the article is by a local writer, phone him or her and try to determine what he or she knows about the subject, or ask the writer to come to class to discuss how information was gathered for the article. Choose five articles and comment on the validity of the impressions they give.

5. Have a person from a local pollution-control agency visit the class and discuss its function.

6. Do an inventory of hazardous chemicals in your home. Determine the proper disposal method for each item found.

7. Draw a diagram tracing the flow of wastewater from your toilet to its discharge into a local body of water.

8. Sit quietly in a room, park, or natural area. Identify the sounds you hear. Rank them as to which are the loudest and which are the most annoying.

9. Keep a detailed diary of the amount of solid waste you produce in one day. List ten actions you could take to reduce the amount of waste you produce.

10. Capture rain or snow and measure its pH.

11. Maintain several bottles of pond water in sunlight. Add different amounts of fertilizer to each bottle and compare how they differ in appearance over time.

Energy Topics

Field Trip Suggestions

1. Visit a power plant: nuclear, coal-fired, wood-fired, or hydroelectric. Determine the quantity of electrical energy produced, the size of the service area, if electricity is sold to other utilities, and whether the facility is used primarily for the base-load or peak-load requirements of the utility. Describe how the utility minimizes its negative ecological impacts.
2. Visit your school's power plant. Determine the amount of energy used per year, and calculate energy use per student per day. What steps has your school taken to reduce energy consumption? Why were they taken?
3. Visit a coal mine, oil field, or gas field. Describe how the company minimizes its negative ecological impacts.
4. Visit an energy information center. (Most power companies provide such services.) Describe five changes you would make and how much energy you would save. What would this mean in monetary terms?
5. Visit a pipeline or powerline right-of-way. Describe three ways the vegetation in the right-of-way differs from the adjacent, less disturbed land.
6. Visit an oil refinery. Describe how the company minimizes its negative ecological impacts. Determine where the company sells its product.
7. Visit a nuclear facility (hospital, X-ray installation, or nuclear power plant). List ten steps taken to assure safety.

Alternative Learning Activities

1. Invite a power company executive to talk to the class.
2. Invite the school's physical plant director to talk to the class about the energy requirements of the school and the costs involved.
3. Trace the path of oil from production facility to gasoline station by interviewing or writing letters to people and asking where they purchase their product (i.e., gas station gets its gas from a distributor, who gets it from a wholesaler, refiner, oil pipeline company, etc.).
4. Trace the path of coal to a power plant, uranium to a nuclear power plant, or natural gas to a home.
5. Draw up a list of ten ways you could reduce energy expenditures. Implement your list.
6. Within your class conduct a contest to see who can reduce energy expenditures the most.
7. Collect all the wastepaper in a particular building or area of your campus and determine how much energy this represents in terms of calories of heat energy.
8. Visit a local hospital to learn how it generates low-level radioactive waste. How does it dispose of this waste? What does it cost?

Policy Issues

Field Trip Suggestions

1. Visit a local zoning board meeting or city council meeting when questions of land-use priorities are on the agenda. List the interest groups present at the meeting and describe the major points of view expressed by each group.
2. Visit a park and have a planner show how decisions were made regarding specific land uses. Select one portion of the park and list the reasons for developing it for its particular use.
3. Visit a supermarket and determine the origin of five fruits or vegetables, two types of fresh meats, one canned meat, one canned fish, one package of coffee, one package of tea, and one package of bread. Some of these items will have the origin printed on the label. To find the place of origin of some of the items, you may need to ask the produce manager or the butcher.
Plot the places of origin on a map of the world. Select one domestic and one foreign item and list the steps necessary to get these items to market.

4. Visit a clothing store and determine the country of origin of five items by reading the labels. Use data from the United Nations Data Books, The Population Reference Bureau Annual Data Sheet, or other sources to determine typical wage rates for people in those countries.
    Plot the countries of origin on a map of the world. Write a paragraph describing why you think these items were manufactured where they were.
5. Visit your state, county, or city officials and discuss environmental legislation with them. Ask them to state their position on a locally important environmental issue. Ask them to list what environmentally significant legislation or ordinances they have supported in the past year.
6. Visit a local recycling center and list the kinds of materials recycled. Describe how the materials are processed within the center. Determine the market for the materials. Who uses the recycled materials, and how much are they willing to pay?
7. Visit a junkyard. Make a list of the kinds of materials accepted. Determine the price paid for each kind of material.
8. Attend a public meeting on an environmental issue.

Alternative Learning Activities

1. Invite your state or federal representatives or senators to address the class or school on environmental issues.
2. Write a letter outlining your position on a piece of environmental legislation to an appropriate government official.
3. Volunteer your time to participate in an environmentally significant activity.
4. Select a piece of land and do an environmental history of it. You may need to interview the current owners and search the records in the county register of deeds office.
5. Select a large piece of land and develop a map that shows suitable uses for various portions of it.
6. Determine what current environmentally significant legislation is before Congress (select one bill). How would passage of the bill affect your area? Keep a log that lists the progress of the bill.
7. Ask a spokesperson of your local water planning board to talk to the class about where water is used and how water use will affect the economy of the area.
8. Use a map of your local community to locate the open space available to the public (playgrounds, parks, golf courses, natural areas, etc.). Do all regions of the community have equal access to open space? How does access to open space relate to economic conditions, politics, and community planning?
9. Identify three examples of good planning and three examples of poor planning in your community. Prepare visual aids that illustrate your examples.
10. Read *The Tragedy of the Commons* by Garrett Hardin and apply his concept to a local situation. Write a paragraph showing how the local situation is related to Hardin's essay.
11. Invite a city planner to discuss how decisions are made relative to the growth of the city.
12. Invite a landscape architect to visit the class and discuss factors that are considered in making a plan.
13. Have a speaker discuss the local business climate.
14. Trace a polystyrene foam container, glass bottle, or plastic bottle from its site of manufacture to its final disposal. (You may need to make several phone calls to get the information you need.)
15. Identify several foreign and domestic foods and drinks. Plot their journey to you on a map of the world. Discuss the importance of these products to the economies of their place of origin. Discuss the roles of politics, energy use, and world hunger as they are related to our use of food.
16. Invite a local corporate leader to discuss corporate policy and environmental issues with the class. Balance this experience by also inviting a local environmental leader to discuss the viewpoint of environmentalists.